T0321149

Adventures in Recreational Mathematics

Volume 1

Problem Solving in Mathematics and Beyond

Print ISSN: 2591-7234
Online ISSN: 2591-7242

Series Editor: Dr. Alfred S. Posamentier
Distinguished Lecturer
New York City College of Technology - City University of New York

---

There are countless applications that would be considered problem solving in mathematics and beyond. One could even argue that most of mathematics in one way or another involves solving problems. However, this series is intended to be of interest to the general audience with the sole purpose of demonstrating the power and beauty of mathematics through clever problem-solving experiences.

Each of the books will be aimed at the general audience, which implies that the writing level will be such that it will not engulfed in technical language — rather the language will be simple everyday language so that the focus can remain on the content and not be distracted by unnecessarily sophiscated language. Again, the primary purpose of this series is to approach the topic of mathematics problem-solving in a most appealing and attractive way in order to win more of the general public to appreciate his most important subject rather than to fear it. At the same time we expect that professionals in the scientific community will also find these books attractive, as they will provide many entertaining surprises for the unsuspecting reader.

Published

Vol. 21 *Adventures in Recreational Mathematics (In 2 Volumes)*
by David Singmaster

Vol. 20 *X Games: Training in Sports to Play in Mathematics*
by Tim Chartier

Vol. 19 *Mathematical Nuggets from Austria: Selected Problems from the Styrian Mid-Secondary School Mathematics Competitions*
by Robert Geretschläger and Gottfried Perz

Vol. 18 *Mathematics Entertainment for the Millions*
by Alfred S Posamentier

For the complete list of volumes in this series, please visit www.worldscientific.com/series/psmb

Problem Solving in
Mathematics and Beyond

Volume 21

Adventures in Recreational Mathematics

Volume 1

David Singmaster
London South Bank University, UK

NEW JERSEY • LONDON • SINGAPORE • BEIJING • SHANGHAI • HONG KONG • TAIPEI • CHENNAI • TOKYO

Published by

World Scientific Publishing Co. Pte. Ltd.
5 Toh Tuck Link, Singapore 596224
USA office: 27 Warren Street, Suite 401-402, Hackensack, NJ 07601
UK office: 57 Shelton Street, Covent Garden, London WC2H 9HE

Library of Congress Cataloging-in-Publication Data
Names: Singmaster, David, author.
Title: Adventures in recreational mathematics / David Singmaster, London South Bank University, UK.
Description: Hackensack, NJ : World Scientific, [2022] | Series: Problem solving in mathematics and beyond, 2591-7234 ; vol. 21 | Includes bibliographical references and index.
Identifiers: LCCN 2020037888 (print) | LCCN 2020037889 (ebook) |
ISBN 9789811225642 (set ; hardcover) | ISBN 9789811226304 (set ; paperback) |
ISBN 9789811226007 (vol. 1 ; hardcover) | ISBN 9789811226502 (vol. 1 ; paperback) |
ISBN 9789811226038 (vol. 2 ; hardcover) | ISBN 9789811226519 (vol. 2 ; paperback) |
ISBN 9789811226014 (vol. 1 ; ebook for institutions) |
ISBN 9789811226021 (vol. 1 ; ebook for individuals) |
ISBN 9789811226045 (vol. 2 ; ebook for institutions) |
ISBN 9789811226052 (vol. 2 ; ebook for individuals)
Subjects: LCSH: Mathematical recreations. | Mathematical recreations--History.
Classification: LCC QA95 .S4958 2022 (print) | LCC QA95 (ebook) | DDC 793.74--dc23
LC record available at https://lccn.loc.gov/2020037888
LC ebook record available at https://lccn.loc.gov/2020037889

British Library Cataloguing-in-Publication Data
A catalogue record for this book is available from the British Library.

For any available supplementary material, please visit
https://www.worldscientific.com/worldscibooks/10.1142/11977#t=suppl

Desk Editors: Vishnu Mohan/Tan Rok Ting

Typeset by Stallion Press
Email: enquiries@stallionpress.com

Printed in Singapore

Contents

Preface

I have always been interested in various mathematical recreations. But this interest became serious after the Rubik's Cube craze in 1978 and onward. The public showed an unprecedented interest in an activity that was only thinly disguised mathematics. I was very much in the midst of this furor when I wrote the first book on the Cube and how to solve it.

As a mathematician, I was aware of the principles the Cube displayed: the symmetries, the power of a sequence of moves to switch pieces of the Cube while not moving most of the rest, called conjugates. Mathematicians had explored the principles a century before; they called it Group Theory. This opened a door wide for me. I began to revisit the many topics in recreational mathematics and pursued their origins. When did the basic puzzles or problems first appear? Who introduced them first? How did they change over time? And how might they inspire today's enthusiasts to take them further? I thought it would be possible to produce a book on the origins of recreational problems.

When I embarked on this, I soon discovered that such information did not exist — for many problems, the origins were completely unknown. I then read numerous works and uncovered early origins of problems in Latin works and in translations of Chinese and Indian works. I began to make notes on early sources. This has grown to a document of about 1000 pages covering 472 topics. I say more about this in the Appendix.

In this process, I have discovered and published material on the origins of numerous problems in various publications. This book is my attempt to gather these results together in one place, in a standard form with many updates. These collective efforts are my own "adventures in recreational mathematics".

Some examples of my adventures are as follows:

- Chapter 4 studies Alcuin, who wrote one of the earliest European mathematical texts, c. 800; it was explicitly a text with recreational exercises. This is not well known. So I undertook to explore the prior scholarship of the Latin text and to annotate Alcuin so the modern reader can easily understand it.
- Chapter 5 studies the Latin text of Abbot Albert, c. 1150; it includes puzzles. I have translated these portions for the modern public, incorporating prior scholarship.
- Chapter 6 presents the work of Pacioli, c. 1500, who published the first book on recreational mathematics. He introduced so many puzzle types that every interested student should know him. However, his output remains hidden in scholarly works. I worked through the various translations and competing ideas to assemble a cogent presentation on Pacioli.

Therefore, much of this volume is not reporting on others' work, but presenting new analyses, translations, and categorizations.

Another aspect of this book is found in Part II, where I have taken up ancient puzzles and found them to be sources of open questions. This explores the "mathematics" of recreational mathematics. Using today's mathematical toolbox, we can ask and answer questions the old masters could only hint at.

A majority of the sources are manuscripts and other documents that are difficult to find and to cite. As a result, the last chapter gives detailed information about these sources, since including it after each chapter would be both lengthy and repetitive. So if a chapter's bibliography has a reference in a font like this, PACIOLI-DVQ, then it indicates that the full citation is in the Appendix.

The companion volume to this is *New Adventures in Recreational Mathematics*. Whereas the emphasis in this book is on the historical roots of these puzzles and problems, the second book mainly concentrates on recently posed problems. In that book not only is the presentation more modern, but also the mathematics is slightly more difficult.

Dedicated to the memory of great puzzlers and friends: Martin Gardner, John Conway, Solomon Golomb, Richard Guy, Edward Hordern and Nob Yoshigahara. My thanks go to Paul J. Campbell for suggestions and to Mike Hammerstone for technical support.

About the Author

David Breyer Singmaster studied at Caltech and received a Ph.D. in mathematics from the University of California-Berkeley in 1966. He taught at the American University of Beirut, later lived in Cyprus, and then came to London in 1970 — and has been based there since. He retired from London South Bank University in 1996 and was designated emeritus in 2020. His interests are in number theory and combinatorics, and the history of mathematics and of science in general.

From 1978 to about 1984, he was the leading expositor of Rubik's Cube. He devised the non-standard notation for it, wrote the first book on the Cube, and later edited Rubik's book into English. Due to revived interest in the Cube, he and some colleagues produced a new book *The Cube: The Ultimate Guide to the World's Bestselling Puzzle — Secrets, Stories, Solutions* in 2009 and *The Handbook of Cubik Math* in 2010.

Since about 1982, he has been working on a history of recreational mathematics, *Sources in Recreational Mathematics*, see Appendix A, which has involved reading and studying mathematics from every culture and period.

He was the opening speaker at the Conference opening the Strens Memorial Collection at Calgary in 1986. He has attended all of the Gatherings for Gardner and spoken at many of the them. He acted as Chairman for four of these. He has attended all of the MathsJam's in England. He was an invited speaker at the Third Iberian Colloquium on Recreational Mathematics in 2012, where he spoke on "Vanishing

Area Puzzles" (Chapter 16). He attended the Fourth Colloquium in Lisbon in 2015, where he spoke on "Early Topological Puzzles" (Chapter 8). His book *Problems for Metagrobologists: A Collection of Puzzles With Real Mathematical, Logical or Scientific Content*, a collection of over 200 problems that he composed since 1988, appeared in 2015.

Chapter 1

What is Recreational Mathematics?

"Les hommes ne sont jamais plus ingénieux que dans l'invention des jeux."
(Men are never more ingenious than in inventing games.)
— Leibniz to De Montmort, 29 July 1715.

"... [I]t is necessary to begin the Instruction of Youth with the Languages and Mathematicks. These should ... be taught to-gether, the Languages and Classicks as ... Business and the Mathematicks as ... Diversion."
— Samuel Johnson, first President of Columbia University, in 1731.

It is worth considering what is meant by recreational mathematics. It is not an oxymoron as many people believe. It is, as the term implies, mathematics which is fun! However, most mathematicians will tell you that their work is fun, even if it is the study of eigenvalues of elliptic partial differential equations. So this definition would encompass all mathematics, which is so general as to be meaningless.[a]

Interested readers of this literature may not be able to solve the problem, but they should be able to appreciate the solution, its relationship to other problems and its cleverness. Some examples can be easily identified as puzzles: get a man with a wolf, a goat and a sack

[a]This chapter appeared in part as: Article 12.3: "Recreational mathematics" in *Companion Encyclopedia of the History and Philosophy of the Mathematical Sciences*, ed. by I. Grattan-Guinness, Routledge, 1993, 1568–1575.

of cabbage across a river where there is a boat which can carry the man and only one of his items. Other examples could be taken as problems: find those integer-sided rectangles whose area is equal to their perimeter. When the solution is too hard, the discussion shifts from recreational to serious, e.g. Fermat's Last Theorem, the Four Color Theorem, the Mandelbrot Set, etc. However, in those cases, the problem remains interesting to the wider public, even though the details are rarefied. We fall into the borderland between Recreational Mathematics and Popular Mathematics.

Hence, there are two criteria — recreational mathematics must be fun and popular! *Popular* means that an interested person should be able to understand the problem. Recreational problems occur in many ways in mathematics. The teaching of mathematics often relies on the fun aspects to make the bridge to mathematics attractive.

In both cases, the fun aspect is often accentuated by posing the problem in a context that is illegal, immoral, or politically incorrect (for one or more reasons), as well as being highly unlikely or even downright impossible. This whimsy is actually important, in that it makes the problem memorable; and the artificiality often eliminates unnecessary complications that tend to occur in reality. Further, the problem may be illustrated or even encapsulated in a physical object that one can see and touch.

The audience is not always the novice. For those striving to learn mathematics, it is often presented as a digression and relaxation. This aspect is already found in the earliest known works of mathematics — the Rhind Papyrus and old Babylonian cuneiform tablets. In these, we see recreational text mixed with instructional text. Such problems also occur in the earliest works of Indian and Chinese mathematics.

Mathematical recreations are as old as mathematics itself. The earliest piece of Egyptian mathematics, the Rhind Papyrus [3] of c. 1800 BCE, has a problem (No. 79) where there are 7 houses, each house has 7 cats, each cat ate 7 mice, each mouse would have eaten 7 ears of spelt (a kind of wheat), and each ear of spelt would produce 7 hekat (a unit of volume) of spelt. Then $7 + 49 + 343 + 2401 + 16807$ is computed. A similar problem of adding powers of 7 occurs in Fibonacci (1202) [4], in a few later medieval texts, and in the children's riddle rhyme "As I was going to St. Ives". Despite the gaps in the history, it is tempting to believe that "St. Ives" is a descendant from the ancient Egyptians. (See Figures 1.1–1.3.)

Figure 1.1. Rhind papyrus [3] No. 79.

Though there is some question as to whether this problem is really a fanciful exercise in summing a geometric progression, it has no connection with other problems in the papyrus and seems to be inserted as a diversion or recreation.

Old Babylonian tablets show fanciful problems that lead to quadratic equations. For example, tablet AO 8862 discusses a field where the length plus the width is known and the area plus the difference of the length and the width is known, see Figure 1.4. This can hardly be considered a practical problem; rather it is a way of presenting two equations in two unknowns, which should make the problem more interesting for the student.

These two aspects of recreational mathematics — the popular and the pedagogic — overlap considerably and there is no clear boundary between them and "serious" mathematics. In addition, there are two other allied fields which contain much recreational mathematics: games and mechanical puzzles.

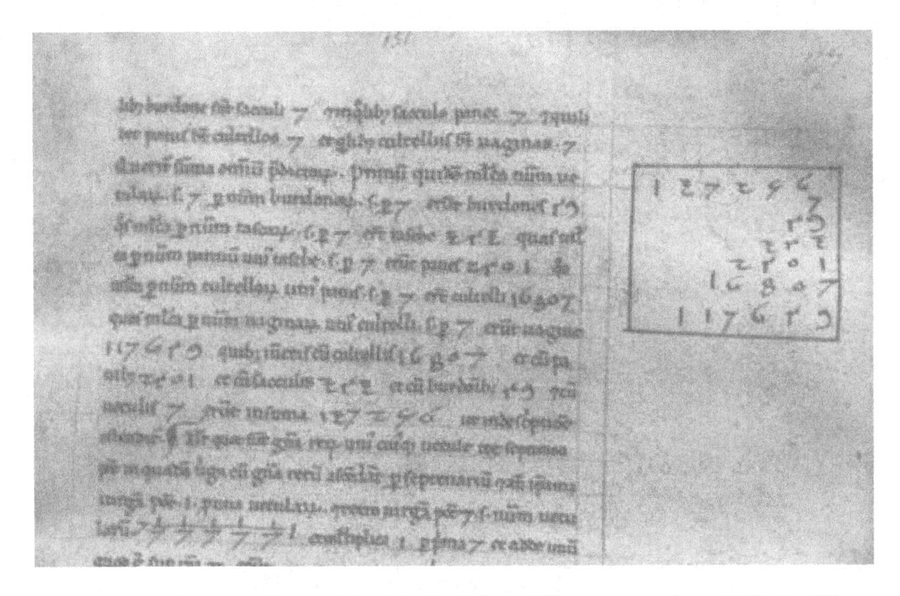

Figure 1.2. Extract from Fibonacci's *Liber Abbaci*, ms. of c. 1275 in Siena, L.IV.20, F. 147r. [2].

Games of chance and games of strategy seem to be about as old as human civilization. The mathematics of games of chance began in the Middle Ages and its development by Fermat and Pascal in the 1650s rapidly led to probability theory and applications. Insurance companies based on this theory were founded in the mid-18th century. The mathematics of games of strategy only started about the beginning of the 20th century, but soon developed into game theory.

Mechanical puzzles range widely in mathematical content. Some only require a certain amount of dexterity; others require ingenuity and logical thought; while others require systematic application of mathematical ideas or patterns, such as Rubik's Cube or the Chinese Rings. The earliest surviving mechanical puzzles seem to be Phoenician puzzle jugs from about 1500 BCE found in Cyprus; such jugs have been around ever since. The Loculus of Archimedes is a set of fourteen pieces which can be assembled into various shapes — an elephant, a boat, etc., like the more recent seven piece Tangrams. The Loculus was known to Archimedes and is mentioned in Classical

Figure 1.3. Postcards illustrating the St. Ives children's riddle rhyme.

literature until the 6th century. Legend ascribes the Chinese Rings to about 200 CE but the earliest records seem to be from the Sung Dynasty (960–1279).

Much magic has a mathematical basis that the magician uses but carefully conceals — e.g. the fact that the opposite faces of a die add up to 7; binary divination; the fact that the period of a perfect (faro or riffle) shuffle of a 52-card pack of cards is 8. See [1].

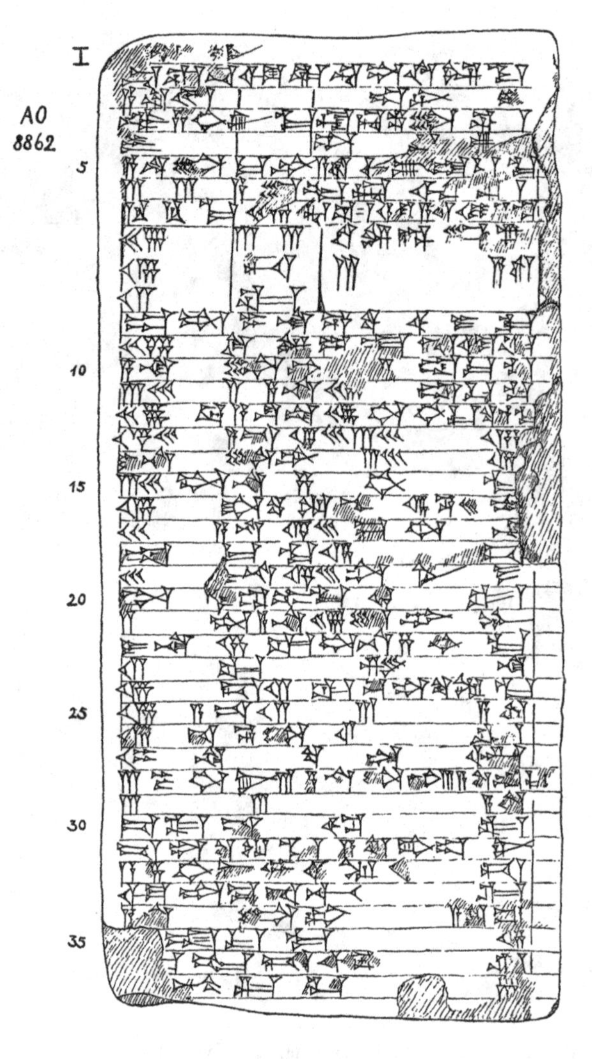

Figure 1.4. Babylonian tablet AO 8862, Face I, in the Louvre. Source: [5], Plate I. "I know the length plus the width of a rectangle is 27, while the area plus the difference of the length and the width is 183. Find the length and the width".

This is a traditional definition of recreational mathematics, but there is scope for further variations. Throughout this book, you will see examples of many types of recreational problems.

Bibliography

[1] P. Diaconis and R. Graham. *Magical Mathematics.* 2012, Princeton University Press, 25–29 & 42–60.

[2] Leonardo Pisano, called Fibonacci. (c. 1170–1240) *Liber Abbaci*, 1202. Translated by Laurence E. Sigler as *Fibonacci's Liber Abaci: A Translation into Modern English of Leonardo Pisano's Book of Calculation.* Springer, New York, 2002.

[3] G. Robins and C. Shute. *The Rhind Mathematical Papyrus, an ancient Egyptian text.* British Museum Publications, 1985.

[4] P. Singh. "The so-called Fibonacci numbers in ancient and medieval India." *Historia Mathematica*, 12 (1985) 229–244.

[5] F. Thureau-Dangin. *Revue d'Assyriologie et d'archéologie orientale*, 29, 1 (1932) 1–10, with four black-and-white plates between pp. 4 and 5.

Part I
Ancient Puzzles

Chapter 2

Puzzles from *The Greek Anthology*

Mathematical problems are found in surprising places. One of the most unlikely places is a collection of Greek poetry known as *The Greek Anthology*. This is a collection of short epigrams, both in verse and in prose, including hymns to the gods, epitaphs, dedications, eulogies, examples of exotic poetic meters, riddles, charades, oracles, poems forming shapes. In addition to these linguistic recreations there were even mathematical problems!

The *Anthology* comes down to us via a unique manuscript in the Palatine Library, Heidelberg, compiled by one Constantine Cephalas in the 10th century. The epigrams date from the 7th century BCE to the 5th century CE. Some are attributed to definite authors, but many are anonymous. Some of the epigrams are known via other sources.

The *Anthology* has been translated into English, both completely and as excerpts. W. R. Paton produced a dual language version in five volumes for the Loeb Classical Library [1], and we will quote from this version. Paton has put all the arithmetical epigrams in Book XIV of volume 5. There are 44 arithmetical problems in Book XIV. Many of these (Paton's 116–146) are specifically attributed to Metrodorus, a grammarian of the 5th century CE. Paton asserts that the other problems (Paton's 1, 4, 6, 7, 11 13, 48, 51) are in the same style and can be also attributed to Metrodorus.

The problems fall into three main types, each with a few minor variants, and a few miscellaneous problems. These are presented in the next section, giving Paton's numbers in parentheses. All the problems are easily solved by a little algebra and many have been used in school books ever since. Answers will be given at the end of the chapter.[a]

It is clear from the problems that most of them were well known before Metrodorus. For example, D. E. Smith traces the cistern problems back to Heron (c. 50 BCE) and traces its many descendants [2].

2.1 The Problems

Diophantos's age

The first problem is famous as it is one of the few references to Diophantos.

> Problem 1 (126). This tomb holds Diophantus. Ah, how great a marvel! the tomb tells scientifically the measure of his life. God granted him to be a boy for the sixth part of his life, and adding a twelfth part to this, he clothed his cheeks with down; He lit him the light of wedlock after a seventh part, and five years after his marriage He granted him a son. Alas! late born wretched child; after attaining the measure of half his father's life, chill Fate took him. After consoling his grief by this science of numbers for four years he ended his life.

The next problem is of the same type and has survived because it refers to Pythagoras. In the *Anthology*, it is attributed to Socrates but Paton presumably includes it in his attribution to Metrodorus.

> Problem 2 (1). Polycrates speaks: Blessed Pythagoras, Heliconian scion of the Muses, answer my question: How many in thy house are engaged in the contest for wisdom performing excellently?
>
> Pythagoras answers: I will tell thee then, Polycrates. Half of them are occupied with belles lettres; a quarter apply themselves to studying immortal nature; a seventh are all intent on silence and the eternal discourse of their hearts. There are also three women, and above the rest is Theano. That is the number of interpreters of the Muses I gather round me.[b]

[a]This appeared in *Mathematical Spectrum* 17:1 (1984/85) 11–15.

[b]Theano is not counted.

There are 19 problems of this form, though some give several numerical values (e.g. the 5 years and the 3 women above). There is another problem of this general type which is a slight variation of the above.

> Problem 3 (143). The father perished in the shoals of the Syrtis, and this, the eldest of the brothers, came back from that voyage with five talents. To me he gave twice two thirds of his share, on our mother he bestowed two eighths of my share, nor did he sin against divine justice.

Several equations

In the next problem, a day is 12 hours.

> Problem 4 (6). "Best of clocks, how much of the day is past?" There remain twice two thirds of what is gone.

There are 8 problems of this form. The next three problems extend the form, the first only slightly.

> Problem 5 (13). We both of us together weigh twenty minae, I, Zethus, and my brother; and if you take the third part of me and the fourth part of Amphion here, you will find it makes six, and you will have found the weight of our mother.

> Problem 6 (51).
>
> (A) I have what the second has and the third of what the third has.
> (B) I have what the third has and the third of what the first has.
> (C) And I have ten minae and the third of what the second has.

> Problem 7 (49). Make me a crown weighing sixty minae, mixing gold and brass, and with them tin and much wrought iron. Let the gold and bronze [brass] together form two thirds, the gold and tin together three fourths, and the gold and iron three fifths. Tell me how much gold you must put in, how much brass, how much tin, and how much iron, so as to make the whole crown weigh sixty minae.

There are two further problems of this basic form but which form a special subtype. This subtype is elsewhere attributed to Euclid, but not in the *Anthology*.

Problem 8 (145).

(A) Give me ten minas and I become three times as much as you.
(B) And if I get the same from you I am five times as much as you.

Cistern problems

These problems and their complications survived as favorites to torture school students well into the 20th century. Recall that a day is 12 hours.

> Problem 9 (135). We three Loves stand here pouring out water for the bath, sending streams into the fair flowing tank. I on the right, from my long-winged feet, fill it full in the sixth part of a day; I on the left, from my jar, fill it in four hours; and I in the middle, from my bow, in just half a day. Tell me in what a short time we should fill it, pouring water from wings, bow, and jar all at once.

> Problem 10 (7). I am a brazen lion; my spouts are my two eyes, my mouth, and the flat of my right foot. My right eye fills a jar in two days, my left eye in three, and my foot in four. My mouth is capable of filling it in six hours; tell me how long all four together will take to fill it.

There are four problems with three spouts and two with four spouts.

There are two similar problems where the individual rates are given differently as in the following.

> Problem 11 (136). Brick makers, I am in a great hurry to erect this house. Today is cloudless, and I do not require many more bricks, but I have all I want but three hundred. Thou alone in one day couldst make as many, but thy son left off working when he had finished two hundred, and thy son in law when he had made two hundred and fifty. Working all together, in how many hours can you make these?

Miscellaneous

There are three remaining uncategorized problems. The latter two have an extra twist. For the first you need to know that one mina = 100 drachms. For the second, you must know there are three Graces.

Problem 12 (12). Croesus the king dedicated six bowls weighing six minae, each one drachm heavier than the other [than the previous one].

Problem 13 (48). The Graces were carrying a basket of apples, and in each was the same number. The nine Muses met them and asked them for apples, and they gave the same number to each Muse, and the nine and three had each of them the same number. Tell me how many they gave and how they all had the same number.

Problem 14 (144).

(A) How heavy is the base I stand on together with myself!
(B) And my base together with myself weighs the same number of talents.
(A) But I alone weigh twice as much as your base.
(B) And I alone weigh three times the weight of yours.

In addition, there are two epigrams, which could be classified as arithmetical but are not really problems.

Epigram 8
The Opposite Pairs of Numbers on A Die.
The numbers on a die run as follows: six one, five two, three four.

Epigram 147
Answer of Homer to Hesiod when he asked the Number of the Greeks who took part in the War against Troy: "There are seven hearths of fierce fire, and in each were fifty spits and fifty joints on them." About each joint were nine hundred Achaeans.

2.2 Solutions and Comments

1. Diophantos lived 84 years.
2. 28.
3. Paton asserts that the elder has $1\frac{5}{7}$ talents, the younger $\frac{22}{7}$ and the mother has 1. However, we get $\frac{15}{8}$, $\frac{5}{2}$ and $\frac{5}{8}$, respectively.
4. $5\frac{1}{7}$ hours are gone.
5. Zethus weighs 12, Amphion 8.
6. A has 45, B has $37\frac{1}{2}$, C has $22\frac{1}{2}$.
7. Gold $30\frac{1}{2}$, brass $9\frac{1}{2}$, tin $14\frac{1}{2}$, iron $5\frac{1}{2}$.
8. $A = 15\frac{5}{7}$, $B = 18\frac{4}{7}$.

9. $\frac{1}{11}$ of a day.
10. $\frac{12}{37}$ of a day $= 3\frac{33}{37}$ hours. Paton remarks that some commentators tried to avoid fractions.
11. $\frac{2}{5}$ of a day.
12. The weights are $97\frac{1}{2}$, $98\frac{1}{2}$, $99\frac{1}{2}$, $100\frac{1}{2}$, $101\frac{1}{2}$, $102\frac{1}{2}$.
13. The twist is that you have not enough data to solve the problem. Any multiple of 12 apples works. If we have $12a$ apples, then each Grace has $4a$ at first and each Grace and Muse has a after sharing.
14. Again there is insufficient information. If we let A be the weight of statue A and a be the weight of its base and let B and b be similarly defined for B, then we have $A + a = B + b$, $A = 2b$, $B = 3a$ and the most we can deduce is $A = 4a$, $B = 3a$, $b = 2a$.

Epigram 147. $315,000 = 7 \times 50 \times 900$.

Bibliography

[1] W. R. Paton. *The Greek Anthology.* Loeb Classical Library, Heinemann, 1916–1918. Volume 5, book XIV, 25–105.
[2] D. E. Smith. *History of Mathematics.* Volume II, Dover, 1958, 532–541.

Chapter 3

Āryabhaṭa and Other Early Indian Mathematicians

This chapter will focus on a number of mathematical recreations with Indian connections.[a] However, the study of early recreations, like early mathematics in general, often raises more questions than it answers. Many such questions will be presented in the hopes that the reader may be able to shed some light on them.

The history of mathematical recreations is a microcosm of the history of mathematics and of the history of culture. Recreational problems are often readily identifiable and hence serve as historical tracers, showing the movement of mathematics (and culture) in time and space. From this study, it has become apparent that modern mathematics has two origins. The geometric and axiomatic approach stems from the Greeks, but the arithmetic/algebraic and practical approach is largely derived from Asiatic sources: Babylonia, China, India and the Arabic world. A number of classic recreations have origins in China, then appear in India, the Arabic world and then Europe, showing a clear process of movement, but sometimes the

[a]Presented to the International Seminar and Colloquium on 1500 Years of Āryabhateeyam, Thiruvananthapuram, 14 Feb 2000 and at the International Seminar on Mathematical Tradition in Kerala, Kanjirapally, 18 Feb 2000. The two lectures were largely disjoint. A version appeared in *Proceedings of The International Seminar and Colloquium on 1500 Years of Āryabhateeyam* [Jan 2000]; Kerala Sastra Sahitya Parishad, Thiruvananthapuram, Kerala, India, 2002, pp. 67–83. Note that there are hardly any images in the early Indian works, so I have used images from medieval western works.

historical record has a gap in one or more of these areas. Other problems originated in Babylonia, then reappear in China, Alexandria and India; some problems originate in India; while some problems appear in widely separated times and places or in only one place. We shall see examples of all of these situations.

Āryabhaṭa's *Āryabhaṭīya* of 499 CE [2] rarely gives numerical examples. Bhāskara I provided the examples in his commentary *Āryabhaṭīya-Bhāṣya* of 629 CE [4] and hence such problems are cited as Āryabhaṭīya-Bhāskara I, 499 CE [4]. These examples are later used, e.g. by Chaturveda Pṛthudakasvâmî in his commentary of 860 CE on the *Bràhma-sphuṭa-siddhânta* of Brahmagupta, 628 CE [6]. Generally, the numerical examples of Chaturveda, as Brahmagupta-Chaturveda, 628 CE [6] are cited. In both cases, the method illustrated by the examples is in the earlier text, but is often so cryptic that one needs the examples to follow it.[b]

Two recreational topics occur so often that it would take a much longer time to discuss them, so we will not look at them here — namely the Chinese Remainder Theorem (which is extensively discussed in many works on Chinese and Indian mathematics) and overtaking and meeting problems (which are very widespread and not very difficult).

3.1 Pythagorean Recreations

The theorem of Pythagoras was certainly known to the Old Babylonians, c. 1800 BCE, and the content of the Babylonian tablet Plimpton 322 is discussed in another chapter. The Babylonians already phrased some Pythagorean problems as recreations. The first type considered is the **Sliding Spear Problem**: a spear (or ladder or pole or beam) of height h is standing up against a wall and when its base is moved out b from the wall, the top descends by a distance d. Hence $b^2 + (h - d)^2 = h^2$. See Figure 3.1.

[b]I have been reading early Indian works on mathematics for about twenty years. I would like to thank all those editors and translators of Indian works, over the last two centuries, who have made early Indian mathematics so accessible to the West.

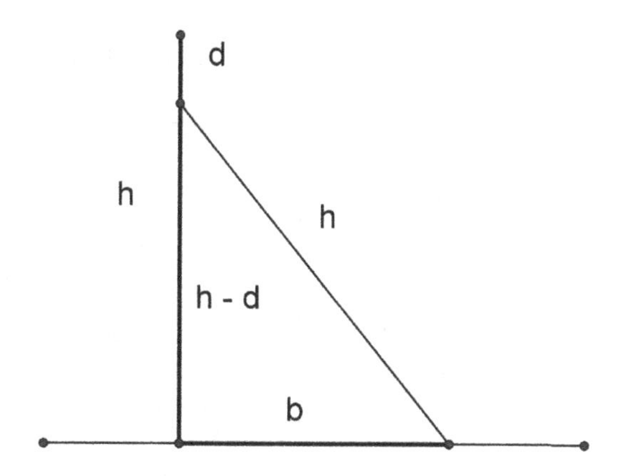

Figure 3.1. Sliding Spear.

Figure 3.2. A later example from *Columbia Algorism* [8], f. 64r. $h = 10$, $b = 6$.

In these problems, two of h, b, d are given and one wants the remaining value. They are denoted, for example, by $h, d = 30, 6$. When h is sought, the h^2 terms cancel and one gets a linear problem. The following early examples have been found:

- Old Babylonian, c. 1800 BCE. Sliding beam, in two versions, with $h, d = 30, 6$ and with $h, b = 30, 18$.
- Seleucid, c. 300 CE. Sliding reed or cane, with $b, d = 9, 3$.

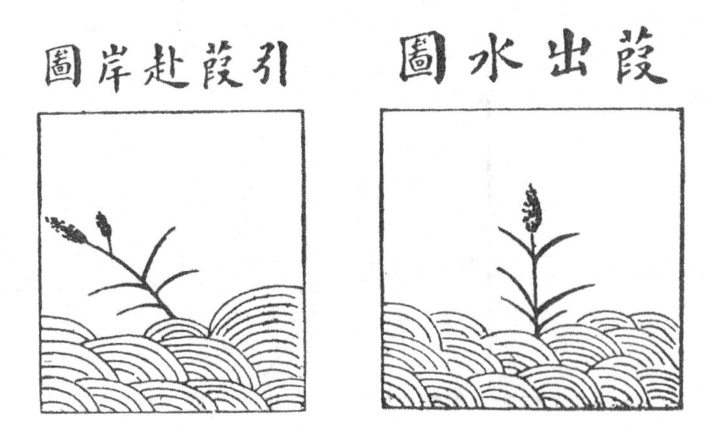

Figure 3.3. From Jiŭ Zhāng Suàn Shú, c 150? [7] (Chap. IX, pp. 92–93, 132–133) $b, d = 5, 1$.

- Ptolemaic, c. 260 CE. Sliding pole, in eight versions, with $h, b = 10, 6$; $14\frac{1}{2}, 10$; $10, 8$; $h, d = 10, 2$; $14\frac{1}{2}, 4$; $10, 4$; $b, d = 6, 2$; $10, 4$.
- Jiŭ Zhāng Suàn Shú, c. 150? [7] (chap. IX, pp. 92–93, 132–133), see Figure 3.3.
  Problem 6. Leaning reed, with $b, d = 5, 1$.
  Problem 7. Version with a rope hanging and then stretched giving $b, d = 8, 3$.
  Problem 8. Leaning ladder, but with vertical and horizontal reversed, with $b, d = 10, 1$.
- *Āryabhaṭa-Bhāskara I*, 499 CE. Leaning lotus. [4] (chap. II, v. 17, part 2, pp. 97–103, 296–300).
  Example 6: $b, d = 24, 8$. Shukla notes that this is used by Chaturveda (see below).
  Example 7: $b, d = 48, 6$.
- Brahmagupta — Chaturveda, 628 CE. Leaning lotus. [6] (chap. XII, section IV, v. 41, pp. 309–310). $b, d = 24, 8$.
- Bhāskara II, 1150 CE. Leaning lotus. [5] (*Lîlâvatî*, chap. VI, v. 151–153, p. 66 = *Bîjagaṇita*, chap. IV, v. 125, p. 204). $b, d = 2, \frac{1}{2}$.

Note that all the Indian examples want to find h and hence are basically linear problems.

It is astounding that Āryabhaṭa and Bhāskara I solve them without using the Theorem of Pythagoras! Instead, these problems are

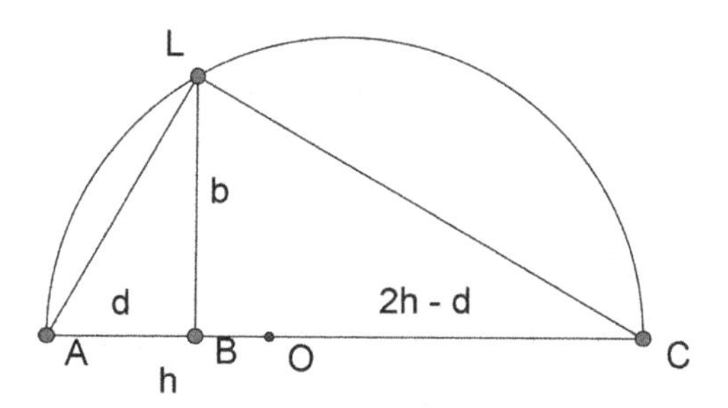

Figure 3.4. Leaning Reed $b^2 + (h - d)^2 = h^2$. Note that O is not the center.

solved by the use of a special case of Euclid III.35, a result easily derived from similar triangles, see Figure 3.4. Let $ABOC$ be a diameter of a circle and let LB be a chord perpendicular to the diameter. Then LAC, BAL, BLC are similar right triangles, so $LB/AB = BC/LB$ or $LB^2 = AB \times BC$.

For the leaning lotus problem, we let LBM be the water level and then the lotus is OBA with O being the root of the lotus at the bottom of the pond. When blown sideways, the lotus is at OL, so $OL = OA = h$, $BL = b$, $AB = d$, $BC = 2h - d$, and our result translates to $b^2 = d(2h-d)$ which is what the Theorem of Pythagoras gives after simplification.

Brahmagupta [8] gives a similar result: by extending LB to form a chord, $(2 \times LB)^2 = 4 \times AB \times BC$ and Chaturveda's example 3 is the same as example 7 of Bhāskara I. Vogel's notes [16] describe several ways to solve this problem, but none of them are like the method of Āryabhaṭa and Bhāskara I [4]. Bhāskara II [5] states that $2h = b^2/d + d$ and develops this from the Theorem of Pythagoras.

The second Pythagorean recreation is the **Broken Bamboo Problem**: A bamboo (or tree) of height h breaks at height x from the ground so that the broken part reaches from the break to the ground at a distance d from the foot of the bamboo. See Figure 3.5. The problem often has a tree beside a river of width d and says the broken part reaches across the river.

As in the common form of the previous problem, the quadratic terms drop out of the solution, leaving a linear problem: $(h - x)^2 = x^2 + d^2$ giving $2hx = h^2 - d^2$, see Figure 3.6. Below early examples of this problem are listed — Babylonian examples of this

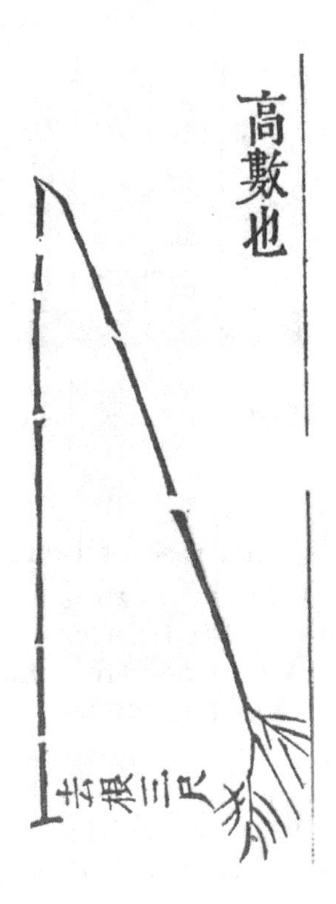

Figure 3.5. Broken Bamboo.

problem have not been seen. In all the cases below, h and d are given and x is sought, so the problem is denoted by (h, d).

- *Jiŭ Zhāng Suàn Shù*, c 150? [7] (chap. IX, problem 13, p. 96). (10, 3).
- Āryabhaṭa-Bhāskara I, 499. [4] (chap. II, v. 17, part 2, pp. 97–103, 296–300).
  Example 4: (18, 6). Shukla notes that this is used by Chaturveda.
  Example 5: (16, 8).
- Brahmagupta — Chaturveda, 628. [6] (chap. XII, section IV, v. 41, p. 309). (18, 6).
- Mahāvīrā, 850. [19] (chap. VII, v. 190–197, pp. 246–248 & 320).
  v. 191. (25, 5), but the answer has $h - x$ rather than x.
  v. 192. (49, 21), but the answer has $h - x$ rather than x.
  v. 193. (50, 20), but with the problem reflected so the known leg is vertical.

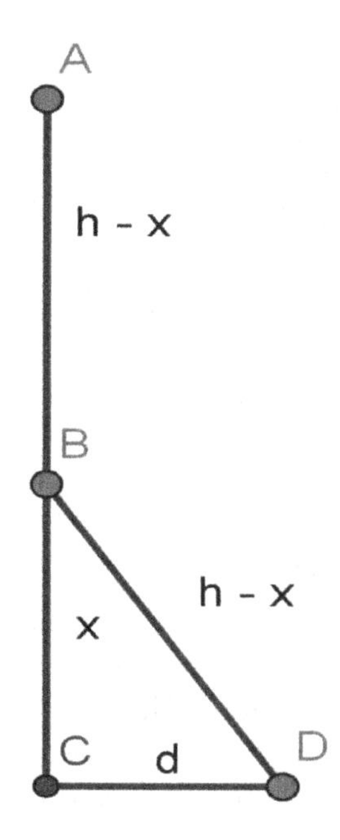

Figure 3.6. From Yang Hui's *Hsiang Chieh Chiu Chang Suan Fa* of c. 1261. Perhaps $h, d = 10, 3$.

v. 196. But this problem imagines two trees of heights H and h, separated by d. The first, taller, tree breaks at height x from the ground and leans over so its top reaches the top of the other tree. If we subtract h from x and H, then $|x - h|$ is the solution of the problem $(H - h, d)$. Because the terms are squared, it does not matter whether x is bigger or smaller than h. He does the case $H, h, d = 23, 5, 12$.

- Bhāskara II, 1150. [5] (*Lîlâvatî*, chap. VI, v. 147–148, pp. 64–65 = *Bîjagaṇita*, Chap. IV, v. 124, pp. 203–204). (32, 16).

Again it is amazing that Āryabhaṭa and Bhāskara I solve it without using the Theorem of Pythagoras! Brahmagupta has his related form and Chaturveda's example 2 is the same as example 4 of Bhāskara I. *Jiŭ Zhāng Suàn Shù* [7] and Mahāvīrā [19] use the Theorem of Pythagoras.

The third type of Pythagorean recreation is the **Two Towers Problem**: there are two towers of heights a, b, which are d apart. The problem has four forms. The most common form is to locate a

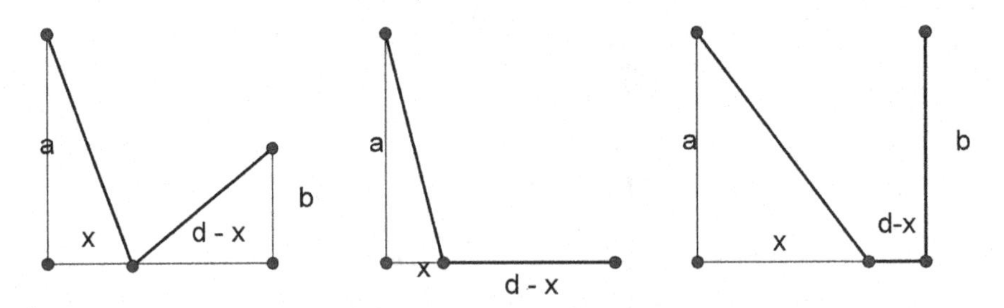

Figure 3.7. Left: Two Towers (Mahāvīrā [19], 850), Center: Hawk & Rat= Peacock & Serpent (Āryabhaṭa I [2], 499) (Bhāskara I [4], 629), Right: Crane & Fish (Āryabhaṭa I [2], 499) (Bhāskara I [4], 629).

well or fountain between the towers so that it is the same distance from the tops of the towers — medieval European versions said that we wanted to have it so that doves take the same time to fly from the tops of the towers to the fountain.

Letting x be the distance from the base of the first tower to the well, we have $a^2 + x^2 = b^2 + (d - x)^2$ and once again the quadratic terms cancel and we have essentially a linear problem. See Figure 3.7. We denote this problem by (a, b, d).

A few years ago, Yvonne Dold [11] pointed out that such problems occur in Mahāvīrā [19] (Chap. VII, v. 201–208, pp. 249–251 & 320). V. 201 gives a general rule and v. 204 gives a straightforward form of (13, 15, 14). However, v. 206 & 208 give a particularly Indian formulation in that we have two hills, upon which "stand two religious mendicants, ..., who can move along the sky," i.e. they behave like the doves of the medieval European versions. These versions have parameters (22, 18, 20), (20, 24, 22). These are the earliest examples of this form of the problem known; no other Indian examples have been seen.

An interesting extension was posed by Cardano — given three towers that are not collinear, find a point on the ground that is equidistant from the tops of the three towers.

However, in Bhāskara I and other Indian works, there was a special case of this problem, of which the earliest is in Āryabhaṭa-Bhāskara I [4], where it involves a hawk and a rat, which we call **Hawk and Rat Problems**. The hawk is sitting on a wall of height a. A rat is distance d away and runs for his hole at the bottom of the wall. The hawk swoops, rather implausibly, at the same speed as the rat can run, and catches the rat at distance x from the wall. This is the same as our two tower problem except that the second tower has no height, i.e. $b = 0$. See Figure 3.7.

This gives us $a^2 + x^2 = (d - x)^2$, which again simplifies to a linear equation. We denote this problem by $(a, 0, d)$. Only the following four examples of this problem have been found, all Indian. The earliest is Āryabhaṭa-Bhāskara I, but it is attributed to unspecified earlier writers.

- Āryabhaṭa-Bhāskara I, 499. [4] (chap. II, v. 17, part 2, pp. 97–103 & 296–300).
  Example 2: Hawk and rat: (12, 0, 24).
  Example 3: Hawk and rat: (18, 0, 81).
- Brahmagupta — Chaturveda, 628. [6] (chap. XII, section IV, v. 41, p. 310).
  Example 4: Cat and rat: (4, 0, 8).
- Bhāskara II, 1150. [5] (*Lîlâvatî*, chap. VI, v. 149–150, pp. 65–66). Peacock and snake: (9, 0, 27). (This occurs in some copies of the *Bîjagaṇita*, between v. 139 and 140 [6] (*Bîjagaṇita*, chap. V, p. 216, note 2).)

Bhāskara I explains the solution in detail; following the analysis of the leaning reed. Brahmagupta-Chaturveda [6] is equivalent to Āryabhaṭa-Bhāskara I [4] (example 2). Looking at Chaturveda, it is seen that turning this sideways gives the same diagram as the broken bamboo problem — the bamboo was BC and breaks at O to touch the ground at L. So the broken bamboo problem (h, d) is the same as the hawk and rat problem $(d, 0, h)$.

Āryabhaṭa-Bhāskara I [3] (examples 8 & 9) gives two examples of another form of the problem, previously unseen. These are **Crane and Fish** problems. A fish is at the NW corner of a rectangular pool of width a and a crane is at the NE corner and they move at the same speeds. The fish swims obliquely to the south side, but the crane has to walk along the edge of the pool. The fish unfortunately gets to the south side, at distance x from the SW corner, just as the crane reaches the same point, and so the fish gets eaten. This again is like our two tower problem, but with the second dove unable to fly, so it has to walk down the tower and across the ground. If we fold down the second tower, this becomes the Hawk and Rat Problem $(a, 0, d + b)$. The resulting equation is $a^2 + x^2 = (b + d - x)^2$. Because the pool

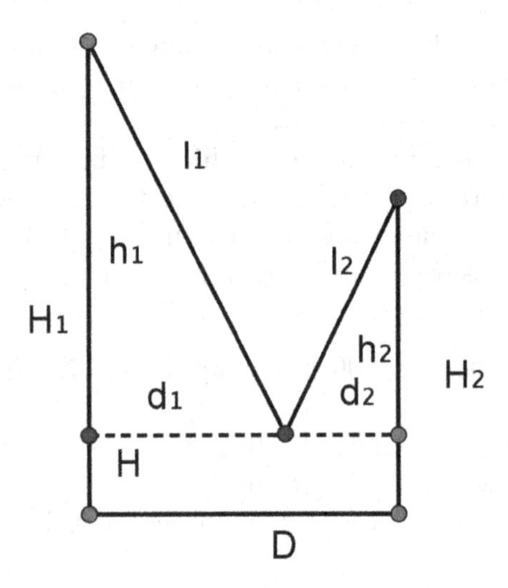

Figure 3.8. Two towers with a weight. The straightforward approach gives seven equations in seven unknowns: $d_1+d_2 = D$, $l_1+l_2 = L$, $h_1+H = H_1$, $h_2+H = H_2$, $d_1^2 + h_1^2 = l_1^2$, $d_2^2 + h_2^2 = l_2^2$, $h_1/d_1 = h_2/d_2$.

is rectangular, the two values a and b are equal, though this is not essential. The two problems have $a, d = 6, 12$; 10, 12.

A final version of the Two Towers Problem first appears in medieval Europe, in dell'Abbaco [1] (problem 158–159, pp. 129–133). Here we have a rope of length L hanging between the tops of the two towers, with a sliding weight on the rope, e.g. attached to a ring or a pulley. His diagram clearly shows the weight in the air, not reaching the ground, so that simple physics tells us the resulting triangles are similar.

Problem 158 has $= H_1, H_2, D = 40, 60, 40$, see Figure 3.8, and a rope of length $L = l_1 + l_2 = 110$, so the rope is more than long enough for the weight to reach the ground, but all he does is "show" that the two parts of the rope are 66 and 44, which is very dubious as there is then slack in the rope, though the diagram clearly shows the weight in the air, see Figure 3.9.

Problem 159 has H_1, $H_2 = 40, 60$ with a rope of length $L = 120$ such that the weight just touches the ground — find the distances of the weight to the towers.

(This inspires a general question not in the original text: when is the rope long enough to reach the ground, and if it does not reach the

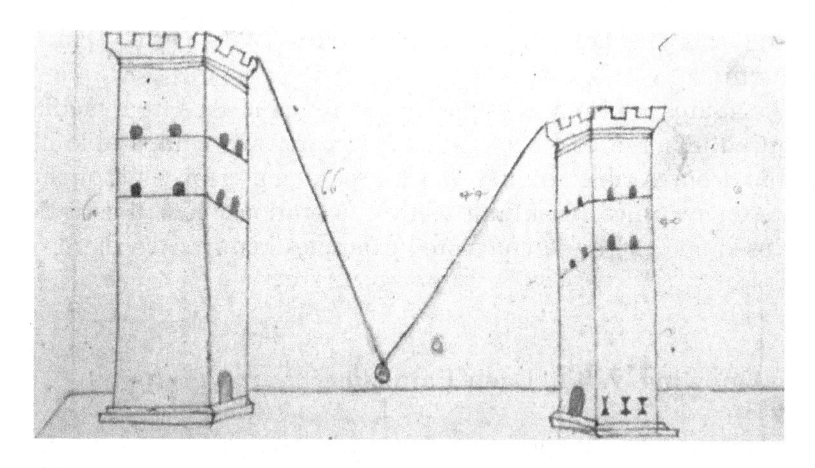

Figure 3.9. dell'Abbaco, c. 1370, [1], problem 158, f. 46r.

ground, how much above the ground is the weight? [20] We will leave you the pleasure of finding this equation. See [21, 22] for a discussion of it. There are two early examples of this problem known, in 1478 and in 1778.)

The final Pythagorean recreation is something which is also found in the early Indian works; no non-Indian examples are known. It is in Brahmagupta — Chaturveda [6] (chap. XII, sect. IV, v. 39, p. 308).

> Two ascetics are at the top of a (vertical!) mountain of height a. One, being a wizard, ascends a distance x straight up and then flies directly to a town which is distance d from the foot of the mountain. The other walks straight down the mountain and to the town. They travel at the same speeds and reach the town at the same time.

There is one example with $a, d = 12, 48$. Bhāskara II [5] (*Lîlâvatî*, chap. VI, v. 154–155, pp. 66–67 = *Bîjagaṇita*, chap. IV, v. 126, pp. 204–205) gives a similar problem with two apes on top of a tower of height a and they move to a point d away, with $a, d = 100, 200$. Algebraically, this is the most complicated of the Pythagorean recreations, giving $a + d = x + \sqrt{(a + x)^2 + d^2}$, but it does again simplify to a linear equation. The authors simply state the result.

We call this the **traveling on the sides of a right triangle problem.**[c]

The figures are from examples of fine illuminations from medieval European manuscripts, because there are not any comparable illustrations from Arabic, Indian or Chinese manuscripts, although we have seen two nice black and white illustrations. Few illustrations were used and there are no colored examples known. Are there such examples?

3.2 Knowing What Each Pair Has

There are two similar problems here, which are the same when we consider just three people, but differ when there are more people. Both types have n people with unknown amounts of money x_i.

In the first type of problem, we know a_i, the total of what all but the ith person have. When there are three people with money x, y, z, this says we know $y + z, x + z, x + y$, i.e. we know what each pair of people has. For four people with money w, x, y, z, the problem tells us $x+y+z$, $w+y+z$, $w+x+z$, $w+x+y$. This is easy to solve, even in general. Let T be the total money present, $T = x_1 + x_2 + \cdots + x_n$. Then we are given $T - x_i = a_i$. Adding all of these together gives us $nT - T = a_1 + \cdots + a_n$, so $T = (a_1 + \cdots + a_n)/(n - 1)$, from which each x_i is determined from $x_i = T - a_i$.

This type of problem is first known in Diophantos, c. 250 [10] (Book I, no. 16 & 17, p. 135). No. 16 is "To find three numbers such that the sums of pairs are given numbers." He does an example with data 20, 30, 40. No. 17 is "To find four numbers such that the sums of all sets of three are given numbers." He does an example with data 22, 24, 27, 20. This problem is closely related to the "Bloom of Thymarides," though it does not appear that the Greeks applied the "bloom" to this type of problem. See [15, 25] for discussions.

The next known appearance of this type of problem is in Āryabhaṭa [3] (chap. II, v. 29, pp. 71 72). He does the problem

[c]These Pythagorean recreations could well be revived for use in schools today. Indeed, I showed some of these in a talk at the First European Congress of Mathematicians in 1990 and someone kindly sent me a copy of the French baccalaureate exam for the next year which had one of these problems on it!

for n people and states the result $T = (a_1 + \cdots + a_n)/(n-1)$. Bhāskara I [5] (chap. II, v. 29, pp. 125–127 & 307–308) gives two examples with the following sets of data: 30, 36, 49, 50 and 28, 27, 26, 25, 24, 23, 21. Perhaps this is an example where Greek mathematics influenced Indian developments, but at some remove. We know other parts of mathematics where there was a clear influence, but the Indian mathematicians made major innovations.

There are examples and a similar problem in the *Bakhshālī* manuscript [3]. This has probably the widest range of dates of any mathematical manuscript, dates ranging from the second century BCE to 12th century CE. Takao Hayashi's 1985 thesis [14] opts for a middling date of c. 7th century, which seems probable. [16] (I 39–42, III 166) [15] (324 & 404–405, art. VI 1) gives an example in the form $x+y=13, y+z=14, z+x=15$, and the other three-person examples mentioned below are in this form. Kaye [16] (I, 39–42) says four more examples "are implied," but the problems containing them are quite varied — see [16] (III, 170–171) [14] (290–291 & 364–366, art. III 4 6). The last of these examples [16] (I, 40, (vi), with the fourth equation omitted; III, 168–169) [15] (288–290 & 364, art. III, 1–3) arises from a different type of problem, discussed below, with five people. This uses an ingenious transformation between knowing the sum of four values and the sum of pairs — i.e. between equations like $v+w+x+y=A$ and $V+X=A$, by letting $V=v+w$, $W=w+x$, $X=x+y, \ldots$.

The next known appearance of this type of problem is in Mahāvīrā [19] (chap. VI, v. 159–162, pp. 136–137 & 315), who uses data: 22, 23, 24, 27.

A related problem arises when we are given what each adjacent pair has, that is, we are given $x_i + x_{i+1} = a_i$. When there are three people with money x, y, z, this says we know $x+y$, $y+z$, $z+x$, and this is the same as the previous type of problem, but with the numbers rearranged. For four people with money w, x, y, z, the problem tells us $w+x$, $x+y$, $y+z$, $z+x$. This problem is remarkable in that it is singular when n is even. For the case of four persons, if we let a, b, c, d be the given values, then we must have $a+c=b+d$ as both are equal to $w+x+y+z$.

The *Bakhshālī* manuscript has an example with five people and data 16, 17, 18, 19, 20 [16] (I, 39–42; III, 166–167) [14] (324–325 & 404–405, art. VI 2–3). Kaye [16] (I, 40, (vii)) asserts that an example with data 36, 42, 48, 54, 60 (or 6, 7, 8, 9, 10) occurs, but "is too mutilated to be sure of".[d]

These problems are also discussed in Datta [9] and Gupta [13]. This type of problem does not reoccur until Fibonacci [12], who considers similar problems and notes that some are inconsistent (e.g. for data 27, 31, 34, 37) and others are indeterminate.

There are several other variations on these problems. Mathematically these are all problems which lead to two equations in two (or three) unknowns (or n equations in n (or $n+1$) unknowns), often formulated as: "If you give me ... then I will have ...." For three or more people, these have two forms, depending on whether "you" means all the other people or just the next person in a cyclic order, but this distinction is ignored here.

The simplest of these problems is the **Ass and Mule Problem**. The ass says to the mule: "If you give me one of your sacks, we will have the same number of sacks." The mule says to the ass: "If you give me one of your sacks, I will have twice as many as you." This was known to the late Greeks and is attributed to Euclid. The only Indian versions are two examples in Mahāvīrā [19] (chap. VI, v. 251–258, pp. 158–159 & 316). This problem will be revisited and discussed at length in Chapter 11.

The next type is related to the above. A **Men Find a Purse Problem** has the first man say: "If I had the purse, I would have a_1 times as much as you.," etc. Here the purse is usually unknown, so the system is indeterminate, but almost all writers are content to obtain the smallest positive integral solution. Again, though known to the late Greeks, the only Indian versions of this are seven examples in Mahāvīrā [19] (chap. VI, v. 233–250, pp. 153–158 & 316), and these introduce several variants of the problem.

The third type is the **Men Buying a Horse Problem** (or a house, etc.), but each man does not have enough money. The first says: "If I had all of what you have, then I could buy the horse", etc. The value of the horse is generally unknown, so this is also an indeterminate problem. This type of problem was known to the Chinese and the Greeks, but the only Indian example is a related problem in the Bakhshālī manuscript [16] (I 39–42,

[d]I have not been able to locate this in his part III or in Hayashi.

Figure 3.10. Pier Maria Calandri [= Benedetto da Firenze], c. 1480, f. 188r.

III 168–169) [14] (288–289 & 364, art. III, 1–3) which leads to equations $T - v/2 = T - w/3 = T - x/4 = T - y/5 = T - z/6 = h$, where $T = v + w + x + y + z$.

There are fewer examples of each type of problem in Indian mathematics than is expected. There may be more unrecognized examples and perhaps readers know of further occurrences in other Indian works.

3.3 The Snail in the Well

A snail (or a serpent, or a dragon, or a lion) is at the bottom of a well D deep. He can climb A during the day, but he slips back B at night. How long does he take to get out?

The "end effect" in this problem is that when the snail gets to within A of the top, the next day he will get to the top and will be out of the well and will not slide back. Earlier writers did not have this clear and simply treated the snail as gaining $v = A - B$ every full day. For example, if the well is 10 deep and the snail goes up 3 every day and back 2 every night, earlier writers would say it took the snail 10 full days to get out, while later writers would

Figure 3.11. dell'Abbaco [= Dagomari] [1], c. 1370, problem 191, f. 53v.

say that after 7 full days, the snail was now only 3 from the top and would get out at the end of the next daytime. The end effect is first known in two works of c. 1370 [8] (problem 67, pp. 88–89) [1] (problem 191, pp. 151–153). It seems that the problem grows out of the usage of unit fractions such as $\frac{1}{2} - \frac{1}{3}$ to specify rates in movement problems. At first this just meant $\frac{1}{6}$ per day, but then it began to be interpreted as going ahead $\frac{1}{2}$ followed by retreating $\frac{1}{3}$, resulting in the end effect. Although this kind of use of unit fractions seems reminiscent of Egyptian mathematics, the earliest examples found are all Indian.

- Āryabhaṭa-Bhāskara I. [4] (chap. II, v. 26–27, pp. 115–122, 304–306, example 4).
  $v = +\frac{1}{2} - \frac{1}{5}, D = 480.$
- Brahmagupta — Chaturveda. [6] (chap. XII, section 1, v. 10, 283–284).
  $v = +\frac{4}{5} - \frac{1}{60}, D = 76,800,000.$

- Bakhshālī manuscript [3].
  A man with wealth 30 earns at rate $\frac{5}{2}$ and spends at rate $\frac{9}{3}$. (Not quite of the type that we are looking for.) [16] (I, 49, III, 216) [15] (334 & 385, art. VIII 3–4).
  Snake entering a hole. $v = +\frac{1}{2} + \frac{1}{18} - \frac{1}{21}$, $D = 360$. [16] (I, 512, III, 222) [14] (346 & 423, art. X 6).
  The last version seems to say

$$v = \frac{\frac{1}{2\cdot 3} + \frac{1}{3} - \frac{1}{4}}{\frac{1}{2\cdot 3}} - \frac{\frac{1}{2\cdot 5}}{\frac{3}{8}}, D = 108,$$

  but the problem is not complete. [16] (I, 51, III 225) [14] (350–351 and 325, art. X 12).
- Mahāvīrā, 850. [19] (chap. V, v. 24–31, pp. 89–90 & 312).
  v. 24. $v = (\frac{1}{5} - \frac{1}{9})/\frac{3}{7}$; $D = 4 \times 99\frac{2}{5}$.
  v. 25. Rate $= \frac{5}{4}/\frac{7}{2} - \frac{5}{32}/\frac{9}{2}$; total $= 70$.
  v. 26. $v = \frac{3}{10}/\frac{11}{2} - \frac{2}{5}/\frac{7}{2}$; $D = \frac{399}{2}$.
  v. 28. A lotus is growing in a well with outflow and evaporation of the water and a turtle is pulling down the lotus; $v = \frac{5}{2}/\frac{3}{3} + \frac{6}{5}/\frac{4}{3} + \frac{5}{2}/\frac{3}{2} - \frac{21}{4}/\frac{7}{2}$; $D = 960$. The problem has 1 for $\frac{4}{3}$, but $\frac{4}{3}$ is needed for the given answer.
  v. 31. A snake is going into a hole while the snake is growing; $v = \frac{15}{2}/\frac{5}{14} - \frac{11}{4}/\frac{1}{4}$; $D = 768$.
- Śrīdharācharya, c. 900, [23] (examples 32-33, pp. 24 & 93) gives some further complex examples.

From this, it seems clear that this type of problem originates in India. We know that it appears in Ṭabarī, c. 1075 [24] (a work which unfortunately has not yet been translated, but is cited by several authorities) and then in Fibonacci, etc., leading up to the development of the end effect in late 14th century Italy. Indeed, dell'Abbaco's discussion gives the older approach and then says that it is wrong and the correct approach is to consider the end effect — "observe and see the tricks which some problems put together." The *Columbia Algorism* [8] considers the end effect and then has a confused statement which says that other writers have failed to see the trick.

We will see in Chapter 4 that Alcuin was in the court of Charlemagne when that court was in contact with Córdoba, Constantinople, and even Baghdad — Haroun al-Rashid sent a white elephant to

Charlemagne in 802 and al-Rashid was also in contact with China. One of Alcuin's problems refers to "a man in the East" and camels, which give us some feeling that he might have obtained these problems from an Eastern source. Although we have no records, it is reasonable to believe that the various embassies between these rulers would have included merchants and administrators with mathematical abilities and interests and they would have exchanged puzzles and stories.

Do any of these Medieval problems have early Indian versions, or versions from anywhere else, which might have been transmitted to Alcuin?

Bibliography

[1] DELL'ABBACO-ARITMETICA.

[2] ĀRYABHAṬA.

[3] BAKHSHĀLĪ.

[4] BHĀSKARA I.

[5] BHĀSKARA II.

[6] BRAHMAGUPTA.

[7] CHIU.

[8] COLUMBIA ALGORISM.

[9] B. Datta. "The Bakhshālī mathematics." *Bull. Calcutta Math. Soc.*, 21 (1929) 1–60. He asserts that the MS is a copy of a commentary on some lost work of the 4th or 5th century.

[10] DIOPHANTOS.

[11] Y. Dold-Samplonius. "Problem of the two towers." in *Itinera Mathematica.* ed. by R. Franci, P. Pagli & L. Toti Rigatelli, 1996, 45–69.

[12] FIBONACCI, pp. 284–286.

[13] R. C. Gupta. "Some equalization problems from the Bakhshālī manuscript." *Indian Journal of the History of Science.* 21 (1986) 51–61.[e]

[14] T. Hayashi. *The Bakhshālī Manuscript An ancient Indian mathematical treatise.* Egbert Forsten, 1995.[f]

[e] He follows Datta in believing that this is a commentary on an early work, though the MS is 9th century as stated by Hoernle.

[f] Based on his 1985 dissertation: A complete edition and translation with extensive discussion of the context of the problems. He dates it as 7th century.

[15] Sir Thomas Little Heath. *A History of Greek Mathematics.* (2 vols.), OUP, 1921, Vol. I, pp. 94–96. (also with corrections, Dover, 1981.)
[16] G. R. Kaye. *The Bakhshālī Manuscript — A Study in Medieval Mathematics. Archaeological Survey of India — New Imperial Series* XLIII: I–III, with parts I & II as one volume, (1927–1933). (facsimile by Cosmo Publications, New Delhi, 1981, pp. 132–133.[g]
[17] Li Yan and Du Shiran. *Chinese Mathematics: A Concise History.* (In Chinese) Commercial Press, Hong Kong, c. 1965.) (Also an English translation by John Crossley & Anthony W. C. Lun. OUP, 1987.)
[18] U. Libbrecht. *Chinese Mathematics in the Thirteenth Century.* MIT Press, 1973.
[19] MAHĀVĪRĀ(CHĀRYA).
[20] D. Singmaster. "Problem 1748" [The two towers]. *Crux Mathematicorum,* 18, 5 (1992) 140; and 19, 4 (1993) 125–127. (David Singmaster, proposer; Dag Jonsson & Hayo Ahlburg, solvers.)
[21] D. Singmaster. "Symmetry saves the solution." in Alfred S. Posamentier & Wolfgang Schulz, eds.; *The Art of Problem Solving: A Resource for the Mathematics Teacher.* Corwin Press, 1996, 273–286.
[22] D. Singmaster. The history of some of Alcuin's *Propositiones.* in: Charlemagne and his Heritage 1200 Years of Civilization and Science in Europe: Vol. 2 Mathematical Arts; ed. by P. L. Butzer, H. Th. Jongen & W. Oberschelp; Brepols, Turnhout, 1998, pp. 11–29.
[23] Śrîdharâchârya. *Pâtîgaṇita.* c. 900. Transcribed and translated by Kripa Shankar Shukla. Lucknow Univ., 1959.[h]
[24] ṬABARĪ.
[25] I. B. Thomas. *Selections Illustrating the History of Greek Mathematics.* Loeb Classical Library, 1939–1941. Vol. 1, pp. 138–141.

[g] This is a rather poor facsimile, indeed Kaye is "Kay" on the dust jacket — but all the text is preserved. He dates it as c. 12th century.

[h] The text is divided into verses and examples, separately numbered by the editor. I will cite the verse (v.) and example (ex.) and the page of the English text.

Chapter 4

Alcuin and his *Propositiones*

The *Propositiones ad Acuendos Juvenes* [Problems to Sharpen the Young] of Alcuin is a major landmark in the history of mathematics in general as well as in the history of recreational mathematics in particular. This chapter will try to explain and justify this thesis. Some of the reasons are general. To the casual observer, the millennium from the fall of Rome to the Renaissance appears as a bleak and indeed Dark Age. Closer observation reveals that major changes begin in Carolingian times — changes which later burst into full development as the Renaissance and the Scientific and Industrial Revolutions. The court of Charlemagne was the center of the revival of learning in Europe. At the same time, an agricultural and technological revolution made it possible to exploit the rich lands of northern Europe. These events are the direct precursors of the Renaissance and the Scientific Revolution. The *Propositiones* appears at this time and place, giving us an excellent view of the intellectual life of this period. The *Propositiones* is in fact the oldest collection of mathematical problems in Latin and really represents the first novel mathematical material to appear in Latin. We shall see that it is a mixture of old and new — there is some appalling ignorance and some ancient problems combined with a truly extraordinary variety of new problems.

More specifically, the *Propositiones* contains 9–11 new types of problems, 3 new variants of problems, and 2 types that are new to Europe. These comprise 24–27 of the 56 problems occurring — a remarkably high percentage of novelty for any collection of this nature and certainly enough to mark it as being of major importance, as shown by the fact that it has been regularly cited in books on recreational mathematics for several hundred years. Consequently, it is very surprising that this manuscript was not critically edited until the 1970s and that the first translations into modern languages only appeared in the 1990s.

Some of the problems have had lengthy developments since the time of Alcuin, occurring in almost every arithmetic/algebra text down to the modern day. One of his River Crossing Problems — that of the Three Jealous Couples — will be described, along with recent work on it, in Chapter 14. However, to appreciate Alcuin's role as a focus of development, we will examine several of the problems which were known before Alcuin. In particular, the Hundred Fowls Problem originated in 5th century China and spread throughout the literate world in the 8th and 9th centuries, occurring in India and the Arabic world at about the same time as it appears in Alcuin — see Chapter 12. Some of the problems are clearly derived from the Roman and Greek traditions, in particular the use of an incorrect Egyptian/Roman formula for the area of a quadrilateral. Some of these appear in Alcuin in novel variants.[a]

4.1 Alcuin

Alcuin (Ealhwine or Alchvine, latinized as Albinus, later nicknamed Flaccus) was one of the most notable intellectuals of 500–1000 CE. He was born in or near York about 732. He was a student, then a teacher and then head (in 778) of the Cathedral School at York, then the most distinguished school in England, perhaps in Europe. St. Peter's School, Clifton, York, claims a continuous history from this time and may even claim to have been founded by Alcuin. See [1] for more details and references to source material.

In 781, Charlemagne met Alcuin at Parma as Alcuin was returning from a mission to Rome. Charlemagne asked Alcuin to direct the school at his capital at Aachen. He arrived in 782 and stayed till 796. He acted as educational advisor to Charlemagne and was responsible

[a]This was presented at the Colloquium Carolus Magnus — 1200 Jahre Wissenschaft in Zentral-Europa, Aachen, March 1995 [18].

for the major educational reform in Charlemagne's empire, which revived learning in Europe. In 789, schools were ordered to be established in every monastery and diocese. At Aachen, he organized and taught at the Palace School, a major instrument of this Carolingian Renaissance, introducing the learning that had been preserved in the northern Anglo-Saxon and Celtic monasteries. Even the King attended. He wrote numerous textbooks that encouraged dialog in teaching. In particular, he wrote on logic and the calendar. "Alcuin sent envoys far and wide to purchase books for his pupils, but the library he gathered must have been small and portable... because the school accompanied the royal household in its wanderings." [10] He may have introduced chess to England. He was slightly involved in the religious and political events of his time, but details are sparse. For example, in 790, he was sent to renew peace with Offa, King of Mercia — Alcuin's native land — leading to the first commercial treaty in English history. Shortly thereafter, he wrote against the Adoption heresy. In 796, he retired from Aachen, but was Abbot of the monastery of St. Martin at Tours. He revived the school there, making it the best in France. He encouraged the production of Bibles and the development of Caroline minuscule script, the ancestor of Roman type. He died at Tours on 19 May 804.

Alcuin is described in the two contemporary biographies of Charlemagne.

> Alcuin "was the most learned man anywhere to be found. Under him the Emperor spent much time and effort in studying rhetoric, dialectic and especially astrology. He applied himself to mathematics and traced the course of the stars with great attention and care. He also tried to learn to write. ..., but although he tried very hard, he had begun too late in life and he made little progress." Einhard — [10] (p. 79).

> Alcuin was "more skilled in all branches of knowledge than any other person of modern times .... The Emperor went so far as to have himself called Alcuin's pupil, and to call Alcuin his master." Notker — [10] (pp. 94–95).

Some of Alcuin's letters to Charlemagne survive. In 798 [5] (pp. 92–93), Alcuin is answering a request to explain "the erratic courses of the planets in the sky." In 799 [5] (p. 93), he comments on

Charlemagne's theories explaining why the moon appeared smaller than calculations predicted on 18 March. Other letters [5] (pp. 93–95) discuss the date of the beginning of the year and the lunar cycle — the calendar had to be adjusted in 798 to make Easter fall on the "right" day.[b]

There was considerable contact between the Court of Charlemagne at Aachen and other courts in Constantinople, Baghdad and Spain. This is of particular interest when we examine the history of some of his problems. Most notably, ambassadors of Haroun al-Rashid landed in Pisa in 801 and in 802, an elephant arrived.

So it is clear that Alcuin was situated at the time and site of a revival of interest in learning and indeed was a principal source of this revival. It is customary to refer to the period from about 300 to about 1200 in Europe as a Dark Age, particularly in science, with only a few stars like Bede and Alcuin shining through the gloom. But a great deal of technological change is now recognized as taking place in northern Europe at this time. The stirrup, mounted knights, the horse harness, heavy plows and horseshoes led to the increased use of horses instead of oxen. This led to a general agricultural revolution in utilizing the fertile heavy soils of northern Europe, converting from a two-stage crop rotation to a three-stage rotation using legumes and hence producing considerably more protein. This resulted in a major increase of population and the improved transportation provided by the horse led to the agglomeration of population in larger towns. This and the improved food supply led to greater specialization, the need for more schooling, etc. Additionally, both wind and water mills began to be introduced. These and related causes led to a major population and economic shift from southern to northern Europe. It seems that this was the first Old World civilization to develop independently of irrigation. See [21] for an exposition of this development.

[b]I was reading through Allott's collection of Alcuin's letters [5] during the Colloquium concert in the Chapel when I found a photograph of the "Throne of Charlemagne" in the Chapel across from me: "We also talked about the pillars which have been set up in the wonderfully beautiful church which your wisdom has prescribed".

So Alcuin's educational reforms occurred in the right context for them to grow from his time onward, along with the technological changes that eventually led to the Renaissance, the Scientific Revolution and the modern world. The increased light that has recently been cast on this era shows that it was not a truly dark age but the forerunner of the modern age which may be dated as starting 1200 years ago.

4.2 The Manuscripts

The *Propositiones* has long been attributed to Alcuin although we have no definite proof. But in 799, Alcuin wrote to Charlemagne: "I have sent you ... some arithmetical curiosities to amuse you on the empty sheet you sent me .... Our friend and helper, [Einhard], ... can also look up the problems in an arithmetic book." It is conjectured that this refers to the *Propositiones*, though we have no trace of any appropriate arithmetic book.[c]

Folkerts has found 13 manuscripts of the text, the earliest being from the late 9th century — 12 in [1] plus another in [2]. Folkerts finds that there are two slightly different forms of the text which had both been printed in Migne's *Patrologia Latina* — one in the works of Alcuin and one in the works of Bede — so it is convenient to refer to these as the Alcuin and Bede texts. (See Appendix A for details of the Migne printings.) The Bede text contains three extra problems, but omits about a third of the answers. The oldest MS is not complete and the next oldest is missing some pages, but both contain the extra problems of the Bede text, which are denoted as 11a, 11b and 33a. The next oldest MS is from the late 10th century and is essentially the Alcuin text. Folkerts says the Bede text represents a poor later edition of the Alcuin, although we do not have any old copies of the Alcuin text. The later text sometimes has modified solutions, although not always for the better. Folkerts produced a critical edition based on 12 MSS in 1978, but it does not vary greatly from the

[c]This is Letter 75, pp. 92–93, in [5]. I have a reference to this as Epistle 101, but none of the other collections cited by Allott have this number for this letter. Allott says it is number 172 in *Monumenta Germaniae Historica* and number 112 in Jaffé.

Alcuin and Bede texts mentioned above [1]. Folkerts & Gericke give both the Latin text and a German translation. However, they give more detailed Latin solutions of several problems than those given in Folkerts — see problems 9, 10, 11, 11a, 11b.

All of these versions conveniently use the same problem numbers.

Alcuin's problems are clearly not original with him. This is particularly noticeable in that he often does not give solutions of the problems — he simply states the answer and then verifies that it works, sometimes making very crude calculations. Some problems have more than one possible answer, but he gives just one answer for these. In most cases he does not show how the answer was found, not even with simple problems solvable by false position. But we do not really know what his sources were. There are some simple problems of types which go back to ancient times and some problems with Greek and Roman origins, but many problems are completely novel and some are new to Europe. It is reasonable to assume that Alcuin got manuscripts from Constantinople, Baghdad and Moorish Spain — or discussed problems with ambassadors from these centers — but no definite evidence survives of such contact and we are left to compare his text with other texts. This will be done below.

As noted above, Alcuin's work in general represents the dawn of the modern age. This is also true of the *Propositiones* — they mark a transition from the old to the new. Surprising as it may seem to other historians, historians of mathematics are well aware that the Roman contribution to mathematics was essentially nothing. Indeed, it has been said the most notable Roman contribution to mathematics was Vitruvius' description of Archimedes' work on Hiero's crown. The Church only encouraged "practical" mathematics — enough Ptolemaic astronomy to compute the date of Easter and some neo-Pythagorean number mysticism to explain the numbers in the Bible — see below under arithmetic progressions for an example.

Alcuin's *Propositiones* is actually the earliest substantial collection of mathematical problems in Latin. By itself, this fact would make it of considerable interest, but the number of novel problems in it makes it extremely surprising that it was not critically edited until the 1970s and that the first translations into modern languages only appeared in the 1990s.

4.3 Managing the Text of *Propositiones*

In researching the history of recreational mathematics, one finds that the *Propositiones* is frequently cited as the earliest problem collection in Latin. The text contains 56 problems, including 9–11 major types of problem which appear for the first time, 2 major types which appear in the West for the first time and 3 novel variations of known problems. By any standard, such a collection is of major historical interest, so it is surprising that a critical edition was not prepared until Folkerts' edition of 1978, and it is even more surprising that no English translation had ever been produced.[d]

After I obtained copies of the Latin text, in 1984 I met John Hadley, a Catholic priest doing an Open University summer school. Hadley kindly translated the text and provided some comments. Folkerts' critical edition had appeared a few years earlier. By 1992 we had annotated the manuscript, combining Folkert's ideas and our own research in the field. We wish to record our thanks to the late Prof. O. A. W. Dilke for comments and corrections. This appeared in the *Mathematical Gazette*.[e] Since then Folkerts & Gericke have presented new ideas. This chapter is an updated and corrected annotated version.

The Alcuin text has 53 numbered problems with answers. The Bede text has 3 extra problems, but the problcms are not numbered, there are only $34\frac{1}{2}$ answers, and there are several transcription errors. Frobenius used the Bede to rectify the Alcuin.

The Alcuin text numbers the problems with Roman numerals; the Bede text has no problem numbers; Folkerts uses Hindu/Arabic numerals. We and Folkerts & Gericke use Hindu/Arabic numerals. The extra problems in the Bede text occur between problems in

[d] I first learned of this collection from Tom O'Beirne's book [16].

[e] The first version of this appeared in *Mathematical Gazette* 76 (475) (March 1992) 102–126. This is an extended and corrected version, retaining a number of details omitted in the published version.

the Alcuin and are numbered by the Alcuin number followed by a, b: 11a, 11b, 33a — as in Folkerts. We give the problem titles in Latin and English, using Folkerts' Latin. The Latin titles almost all begin "problem about" ("Propositio de") but we will omit the words "Propositio de" and "problem about" in order to make the titles briefer. In the actual problems, Alcuin, Folkerts, and Folkerts & Gericke use Roman numerals for numbers, but the Bede uses Hindu/Arabic numerals. For everyone's convenience, we and Folkerts & Gericke use the Hindu/Arabic numerals in our translations. However, many of the numbers are actually given in words and we have then generally retained the use of words as a characteristic of the original. The last sentence of each problem usually starts with a formula like *Dicat, qui potest* (Speak, who can) which has been dropped from the translation. Jens Høyrup translates this as "Tell me, whoever is worth anything."

The Alcuin text has solutions directly after the problems, while the Bede gives all his $34\frac{1}{2}$ solutions together at the end of the problems. Folkerts sometimes omits solutions that occur in the Alcuin or Bede text, indicating that the solution is probably not original, but Folkerts & Gericke usually give solutions for these. Folkerts and Folkerts & Gericke sometimes give two solutions when the two groups of texts differ — see 22, 25, 29, 30, 31 — the second solution being from the second group. Throughout, the solutions are generally very sketchy, usually simply verifying that the answer works, with no indication of how it was derived.

Units: The most common units of money at the time were the libra, solidus, and denarius which are the originals of the English system of the pound, shilling, and (old) penny, so we will use the old English names. 12 pence = 1 shilling, 20 shillings = 1 pound. See problem 1 for *leuca* and pace. See problem 9 for cubit. See problem 22 for perches and acres. The units of weight were the *libra* and *uncia*, the originals of the pound and ounce, except that there were 12 ounces to the pound. We shall emphasize this by referring to troy ounces the first time we use ounces in a problem. Units of volume vary. See problems 8, 32, and 50 for *metreta, metrum, modius* or *modium, sextarius* and *merus*. Modius and metrum are often translated as "measure" when the size is not important. (For more information on Latin units, see [17].)

4.4 An Annotated Translation of *Propositiones*

Here Begin the Problems to Sharpen the Young
Incipiunt Propositiones ad Acuendos Juvenes

1. A slug — *De limace.*

A leech[f] invited a slug for lunch a leuca away. But he could only walk an inch a day. How many days will he have to walk for his meal?

SOLUTION. A leuca is 1500 paces, that is 7500 feet or 90,000 inches. It will take him as many days as inches, which makes 246 years and 210 days.

NOTES. Here and in Problem 52, Alcuin uses a *leuca* as a measure of length. In classical times, it was defined as 1500 paces, while a modern league is 3000 paces. Folkerts spells it *leuva*, but the Alcuin and Bede texts give *leuca* and some MSS give leuga. These are double paces, what we might call two steps. Such a pace was reckoned at five feet. Hence a leuca was little under a mile and a half, while a league was a little under three miles. In fact, our mile derives from the Roman *mille passus*, indicating a thousand paces, so perhaps a pace should be considered as 5.28 feet, a league as three miles and a leuca as a mile and a half.

2. A walker — *De viro ambulante in via.*

A man walking along a road saw others coming towards him, and he said to them: "I wish there were others there with you, as many as you are, plus a quarter of the sum that would be, plus half of that last amount. Then with me as well, there would be 100 altogether." How many did he see on the road?

SOLUTION. 36. The same again makes 72; a quarter of this is 18, and half of 18 is 9; 72 and 18 is 90; add 9 to make 99; and with the speaker, that makes 100.

NOTES. The Latin has "a half of a half" for quarter. This is a straightforward problem of a type which already occurs in the Rhind Papyrus, where they are known as *aha* or "heap" problems.

---

[f] The Latin is *hirundine* which is swallow, though *hirudo* (leech) is more like a slug and Hadley gave leech.

The Greek problem of "Diophantos' Age" is the same. In the Middle Ages, numerous types appeared, with this type being called a "God Greet You Problem", since that was the usual opening salutation between the travelers.

3. Two travelers and a flock of storks — *De duobus proficiscentibus visis ciconiis.*

Two walkers saw some storks and wondered how many there were. Conferring, they decided: if there were the same number again, and again, and then half of a third of the sum that would make, plus two more, there would be 100. How many storks were seen?

SOLUTION. 28. For 28 and 28 and 28 make 84. Half of a third of 84 is 14, which added to 84 makes 98. Add 2 and there appears 100.

4. Some horses grazing — *De homine et equis in campo pascentibus.*

A man saw some horses at pasture and wished that they were his, and that there were others with them that were his, the same number again, plus a quarter of the sum that would result, for then he would glory in 100 horses. How many did the man see at pasture?

SOLUTION. There were 40 horses grazing. The same number again would make 80; a quarter of that total is 20, which added to 80 makes 100.

5. A merchant and 100 pence — *De emptore in C dinariis.*

A merchant wanted to buy a hundred pigs for a hundred pence. For a boar, he would pay 10 pence; and for a sow 5 pence; while he would pay one penny for a couple of piglets. How many boars, sows, and piglets must there have been for him to have paid exactly 100 pence for the 100 animals?

SOLUTION. Take 9 sows and a boar, to the value of 55 pence altogether, and 80 piglets at 40 pence, which brings the total of pigs to 90. For the other 5 pence, take 10 piglets, and that brings the total of pigs and of pence to 100.

NOTES. This is the first European appearance of this type of problem, usually known as "The Hundred Fowls" because the earliest version, in 5th century China, involves 100 fowls for 100 coins. (Several other names were common in the Middle Ages.) See Chapter 12 for a full discussion. By the time of Alcuin, the problem was well known throughout the world — it appears in Indian and Arabic texts of the 9th century with several problems and variations, e.g. with four or five types of animals and with all solutions sometimes given or described. Seven (or eight) problems of this type appear in the *Propositiones*, with an additional one in the Bede text: 5, 32, 33, 33a, 34, 38, 39, 47. (53 may be a corruption of this type of problem.) The answers are simply stated without any indication of their derivation or of the possibility of other answers. All of the Alcuin problems, except 34, have only one answer with positive values, but 38, 39 and 47 admit a second solution with one value being zero. Problem 34 has 6 positive answers, but Alcuin gives only one — there is an extra answer with a zero value. The extra Bede Problem 33a has 5 answers with positive values, but only one is given — there are two extra solutions with a zero value. It is easy to solve these problems with algebraic notation.

6. Two wholesalers with 100 shillings — *De duobus negotiatoribus C solidos communis habentibus.*

Two wholesalers with 100 shillings between them bought some pigs with the money. They bought 5 pigs for every two shillings, intending to fatten them up and sell them again at a profit. But when they found that it was not the right time of year for fattening pigs and they were not able to feed them through the winter, they tried to sell them again to make a profit. But they could not, because they could only sell them for the price they had paid for them — two shillings for each 5 pigs. When they saw this, they said to each other: "Let's divide them." By dividing them, and selling them at the rate they had bought them for, they made a profit. How many pigs were there, and how could they be divided to make a profit, which could not be made by selling them all at once?

Solution. Firstly, there were 250 pigs bought with 100 shillings at the above-mentioned rate, for five 50s are 250. On division, each merchant had 125. One sold the poorer quality pigs at three for a shilling; the other sold the better quality at two for a shilling. The one who sold the poorer pigs received 40 shillings for 120 pigs; the one who sold the better quality received 60 shillings for 120 pigs. Then there remained 5 of each sort of pig, from which they could make a profit of 4 shillings and 2 pence.

NOTES. The final profit is obtained by selling the remaining pigs at the same prices, yielding 1 shilling and 8 pence plus 2 shillings and 6 pence. This is the first known appearance of this problem, variously known as "Applesellers' Problem", "The Missing Penny", or "The Marketwomen's Problem". The basis of the problem is that one is erroneously averaging equal numbers at 2 for a shilling and at 3 for a shilling to get an "average" price of 5 for two shillings. After Alcuin, this problem appears in almost every European problem book — a non-European version does not seem to exist.

7. A dish weighing 30 pounds — *De disco pensante libras XXX.*

A dish, weighing 30 pounds, is made of gold, silver, brass and lead. It contains three times as much silver as gold; three times as much brass as silver; and three times as much lead as brass. How much is there of each?

SOLUTION. There are nine troy ounces of gold; three times 9 ounces of silver, that is, 2 pounds 3 ounces of silver; three times as much brass, that is, 6 pounds 9 ounces of brass; and three times 6 pounds 9 ounces of lead, that is, 20 pounds 3 ounces of lead. 9 ounces, plus 2 pounds 3 ounces, plus 6 pounds 9 ounces, plus 20 pounds 3 ounces, makes 30 pounds.

NOTES. Alcuin gives a second solution, using *solidi* [silver] (shillings), of which there are 20 to the pound. This gives: 15, 45, 135, 405, making 600 shillings in all. Folkerts & Gericke [2] gives some discussion on geometric progressions but admit this problem is not using the idea.

8. A cask and three pipes — *De cupa.*

A cask is filled with 100 *metretae* through three pipes. One third plus a sixth of the capacity flows in through one pipe; one third of the capacity flows in through another pipe; but only one sixth of the capacity flows in through the third pipe. How many *sextarii* flow in through each pipe?

SOLUTION. 3600 *sextarii* flow in through the first pipe; 2400 through the second pipe; and 1200 through the third.

NOTES. A *metreta* was about 9 gallons. A *sextarius* was about a pint. There were 72 *sextarii* to the *metreta*. Many authors, e.g. [2]

(p. 361), have cited this as a "Cistern Problem", but it does not have the characteristic usage of rates that occurs in such problems.

9. A cloak — *De sago.*

I have a cloak, 100 cubits long and 80 cubits wide. I want to make small cloaks with it, each small cloak 5 cubits long and 4 wide. How many small cloaks can I make?

SOLUTION. John Hadley says: "400. The explanation given for this answer was less than enlightening." He conjectures that a line has been omitted by a scribe. [2] (p. 30) discusses versions in the MSS. The Bede text can be translated as follows.

Solution 1. The eightieth part of 400 is 5 and the hundredth part of 400 is 4. Both eighty 5s and one hundred 4s give the same result, 400. There are that many cloaks.

Solution 2. Folkerts & Gericke [2] gives a better Latin text, which translates as follows:

> Since each cloak has length 5 cubits and width 4, take the fifth part of 100 and get 20, and the fourth of 60 makes similarly 20. Hence make twenty twenties which gives 400. That many cloaks of length 5 and width 4 cubits can be made.

NOTES. A cubit is about half a yard. The Latin has LX (60) instead of LXXX in other versions.

10. A linen problem — *De linteo.*

I have a piece of linen, 60 cubits long and 40 wide. I want to cut it into portions, each of which will be 6 cubits long and 4 wide. How many portions can be made from it?

SOLUTION. A tenth part of sixty is 6; a tenth of forty is 4. Since we have a tenth of sixty and a tenth of forty, we find we have 100 portions 6 cubits long and 4 cubits wide.

NOTE. Folkerts & Gericke [2] give a more detailed Latin version; the changes are not significant.

11. Two men marrying each other's sister — *De duobus hominibus singulas sorores accipientibus.*

If two men each take the other's sister in marriage, what is the relationship of their sons?

SOLUTION. The Alcuin text has a solution, but the Bede text does not. Folkerts does not give a solution in his edition, which indicates that the solution does not seem to be authentic, though he quotes two solutions from MSS; Folkerts & Gericke [2] combines these.

> Solution 1. If I take the sister of my friend and he [takes] mine, and children are born to us, then I am uncle to the sons of my sister and she is aunt to my sons, and that is the relationship between them.[g]
>
> Solution 2. Therefore my son and the son of my sister are called cousins of some sort.

NOTE. This problem and the Bede problems 11a and 11b are the earliest known examples of what I call "Strange Families Problems". Folkerts [1] (p. 38) says one gives up trying to answer these!

11a. Two men marrying each other's mother — *De duobus hominibus singulas matres accipientibus.*

If two men each take the other's mother in marriage, what would be the relationship between their sons?

SOLUTION. Folkerts [1] again does not give a solution, but quotes a solution added to one MS: "Hence my son and the son of my mother are uncles and nephews." This is given in [2].

NOTE. This is the situation which leads to "I'm my own Grandfather." It has been popular ever since Alcuin and I have five references to reported occurrences of the actual situation.

11b. A father and son and a widow and her daughter — *De patre et filio et vidua eiusque filia.*

If a widow and her daughter take a father and son in marriage, so that the son marries the mother and the father the daughter, what is the relationship of their sons?

SOLUTION. Again Folkerts [1] gives no definitive solution, but the same MS mentioned in 11a gives: "Hence my son and the son of my father are uncle and nephew of one another." Again this is given in [2].

[g] The question seems to be asking for the relation between the sons, but the solution does not answer this.

12. A father and his three sons — *De quodam patrefamilias et tribus filiis eius.*

A father, when dying, gave to his sons 30 glass flasks, of which 10 were full of oil, 10 were half-full, and the last 10 were empty. Divide the oil and the flasks so that each of the three sons receive equally of both glass and oil.

SOLUTION. There are three sons and 30 flasks. Of the flasks, 10 are full and 10 are half-full and 10 are empty. Three tens are 30. To each son will come ten flasks as his portion. But divide them as follows: to the first son give the 10 half-full flasks; then to the second give 5 full and 5 empty flasks; and similarly to the third; and there will be equal division among the three sons of both oil and glass.

NOTES. This is the earliest known example of the "Barrel Sharing Problem", which is discussed in Chapter 15. They were popular in Europe after Alcuin, but I know of no non-European examples. There are five solutions, which can be easily found. A solution is determined by the distribution of the full flasks, which can be as follows: 5 5 0; 5 4 1; 5 3 2; 4 4 2; 4 3 3.

13. A king and his army — *De rege et de eius exercitu.*

A king ordered his servant to collect an army from 30 manors in such a way that from each manor he would take the same number of men as he had collected up to then. The servant went to the first manor alone; to the second he went with another; to the next, three went. How many were collected from the 30 manors?

SOLUTION. After the first stop, there were 2 men; after the second 4; after the third 8; after the fourth 16; after the fifth 32; after the sixth 64; after the seventh 128; after the eighth 256; after the ninth 512; after the tenth 1024; after the eleventh 2048; after the twelfth 4096; after the thirteenth 8192; after the fourteenth 16,384; after the fifteenth 32,768; after the sixteenth 65,536; after the seventeenth 131,072; after the eighteenth 262,144; after the nineteenth 524,288; after the twentieth 1,048,576; after the twenty-first 2,097,152; after the twenty-second 4,194,304; after the twenty-third 8,388,608; after the twenty-fourth 16,777,216; after the twenty-fifth 33,554,432; after the twenty-sixth 67,108,864; after the twenty-seventh 134,217,728; after the twenty-eighth 268,435,456; after the twenty-ninth 536,870,912; after the thirtieth 1,073,741,824.

NOTES. The intent is that three others go with him, but the Latin says three come to the third farm. Folkerts & Gericke [2] observe that one 15th century MS correctly has 4 here. Migne stops at the fifteenth value since the values were incorrectly given in the various MSS that he saw and since they are easy enough to obtain by always doubling the previous number. Folkerts [1] gives all the values. Euclid already gives the rule for summing a geometric progression but this problem has not been treated as a sum since we have $1 + 1 = 2$; $1 + 1 + 2 = 4$; $1 + 1 + 2 + 4 = 8$; $\ldots$, so the totals are themselves a geometric progression.

14. An ox — *De bove.*

An ox plows a field all day. How many footprints does he leave in the last furrow?

SOLUTION. An ox leaves no trace in the last furrow, because he precedes the plow. However many footprints he makes in the earth as he goes forward, the cultivating plow destroys them all as it follows. Thus no footprint is revealed in the last furrow.

15. A man plowing — *De homine.*

How many furrows has a man made in his field, when he has made three turnings at each end of the field?

SOLUTION. At each side of the field he has made three turnings, making seven furrows.

NOTE. The Alcuin text gives 6; Bede, Folkerts and Folkerts & Gericke give 7.

16. Two men leading oxen — *De duobus hominibus boves ducentibus.*

Two men were leading oxen along a road, and one said to the other: "Give me two oxen, and I'll have as many as you have." Then the other said: "Now you give me two oxen, and I'll have double the number you have." How many oxen were there, and how many did each have?

SOLUTION. The one who asked for two oxen to be given him had 4, and the one who was asked had 8. The latter gave two oxen to the one who requested them, and each then had 6. The one who had first received now gave back two oxen to the other who had 6 and so now had 8, which is twice 4, and the other was left with 4, which is half of 8.

NOTES. This is usually called "The Ass and Mule Problem" since the earliest known version, in Greek verse attributed to Euclid, has an ass and a mule carrying sacks. Diophantos already gives a general solution of such problems and several examples. Two versions occur in Metrodorus' problems in the *Greek Anthology* [15] and it also appears in Mahāvīrā (c. 850) [14] and al Karagi (c. 1010) [3]. However, Alcuin's version is unique in that the second statement is considered as happening after the first statement is actually executed. The usual form has the second person saying: "But if you give me ...", so with our numbers, the solution would then be 10, 14 since $10 + 2 = 14 - 2$ and $14 + 2 = 2 \times (10 - 2)$. Folkerts & Gericke discuss an example from the early Chinese text, the *Chiu Chang Suan Ching* [8], but this is an example of "Men Buying a Horse", which is a related, but not identical, problem. The general question is discussed and generalized in Chapter 11.

17. Three friends and their sisters — *De tribus fratribus singulas habentibus sorores.*

Three friends, each with a sister, needed to cross a river. Each one of them desired the sister of another. At the river, they found only a small boat, in which no more than two of them could cross at a time. How did they cross the river without any of the women being defiled by the men?

SOLUTION. First of all, I and my sister would go into the boat and travel across; then I'd send my sister out of the boat and I would cross the river again. Then the sisters, who had stayed on the bank, would get into the boat. These having reached the other bank and disembarked, my sister would get into the boat and bring it back to us. She having got out of the boat, the other two men would board and go across. Then one of them with his sister would come back to us in the boat. Then I and the man who had just crossed would go over again, leaving our sisters behind. Having reached the other side, one of the two women [If you are keeping track carefully, you will see that there is only one woman on the far bank at this point!] would take the boat across, and having picked up my sister, would come back to us. Then he whose sister remained on the other side would cross the river in the boat and bring her back with him. That would complete the crossing without anything untoward happening.

NOTES. The Latin words are *frater* and *soror*, which can mean "friend" as well as "brother" or "sister". Three male friends, each with a sister, seems the most likely meaning. Folkerts [1] (p. 38) refers to three men and their wives. The meaning is unclear [2] and translates literally as three brothers, each with a sister. The German for "brother" can also mean "friend" or "friar". Perhaps we should have three monks and three nuns?

This and the next three problems are the earliest known appearances of "River Crossing Problems" giving the three classic types. In recent years, versions have been reported from Africa, but no old non-European sources are known. This problem is usually called "The Jealous Husbands" or "The Jealous Couples" and is discussed in Chapter 14.

The exact rules of behaviour are not made explicit in this problem, but the traditional interpretation is that no man is willing to allow another man to be with his sister unless he is present. Tartaglia claimed that he could get four couples across, but at one point he has a husband take his wife across to leave her with the other wives, but this assumes that the man does not get out of the boat and attack the women on the shore. Other authors have a more pessimistic attitude toward human nature and do not permit such a situation, in which case four (or more) couples cannot cross and Alcuin's solution for three couples in 11 crossings is minimal, though the crossings can be somewhat varied.

18. A wolf, a goat and a bunch of cabbages — *De lupo et capra et fasciculo cauli.*

A man had to take a wolf, a goat and a bunch of cabbages across a river. The only boat he could find could only take two of them at a time. But he had been ordered to transfer all of these to the other side in good condition. How could this be done?

SOLUTION. I would take the goat and leave the wolf and the cabbage. Then I would return and take the wolf across. Having put the wolf on the other side, I would take the goat back. Having left that behind, I would take the cabbage across. I would then row across again and having picked up the goat, take it over once more. By this procedure, there would be some healthy rowing, but without any lacerating catastrophe.

NOTES. There is an alternative solution which is the given solution viewed backward. The objects being transported have been varied in many ways, including a fox, duck and some grain, as well as the African leopard, goat and cassava leaves. One can extend this to more objects: A, B, C, D, ..., where each object is incompatible with its two neighbors, but then one of the objects has to be able to row.

19. A very heavy man and woman — *De viro et muliere ponderantibus plaustrum.*

A man and woman, each the weight of a cartload, with two children who together weigh as much as a cartload, have to cross a river. They find a boat which can take only one cartload. Make the transfer, if you can, without sinking the boat.

SOLUTION. First, the two children get into the boat and cross the river; one of them brings the boat back. The mother crosses in the boat; and her child brings the boat back. His brother joins him in the boat and they go across; and again one of them takes the boat back to his father. The father crosses; and his son, who had previously crossed, having boarded, returns to his brother; and both cross again. With such ingenious rowing, the navigation may be completed without shipwreck.

NOTES. Cartload literally is the weight of a loaded cart. I am unsure if this includes the cart or not, but the latter seems more reasonable. A variation of the problem has the couple meeting unrelated children at the shore and the children want to finish on their original side with their boat.

20. Hedgehogs — *De ericiis.*

The Latin text seems to be defective. Hadley did not try to translate it. It seems to say: "About a male and female hedgehog with two young, having weight, wanting to cross a river."

SOLUTION. Similarly to the above, first send across the two infants and one of them returns in the boat. The father entering it, crosses over; and the infant who first crossed with his brother, returns to the bank. His brother having gotten in again, both cross over; one of them gets out and the other returns the boat to the mother. The mother enters and crosses over. Then that son, who had previously crossed with his brother, again gets in and returns the boat to his brother. Both being in the boat, they come over, and complete the crossing without shipwreck.

NOTE. The reader will notice that this is essentially the same problem as Problem 19.

21. Sheep in a field — *De campo et ovibus in eo locandis.*

There is a field 200 feet long and 100 feet wide. I want to put sheep in it, so that each sheep has five feet by four. How many sheep can I put in there?

SOLUTION. This field is 200 feet long and 100 feet wide. The number of fives in 200 is 40; dividing 100 by 4, the fourth part of 100 is 25. Either 40 twenty-fives or 25 forties makes a thousand. This is therefore the number of sheep that can be put into the field.

NOTES. Problems 21–25 and 27–31 are all similar. Folkerts [1] (pp. 39–41) notes that these problems are identical to IV: 30-39 in *Geometria incerti auctoris* [12], an anonymous 10th century geometry MS, related to Gerbert's work. It was long attributed to Gerbert, but is now considered to predate him. Folkerts compares the problems and solutions and concludes that Alcuin and *Geometria incerti auctoris* are independent versions, presumably from a common source, but *Geometria* is a considerably better version. Following [2], I will denote this as *GIA* [12] and will add some of their comments about *GIA*.

22. An irregular field — *De campo fastigioso.*

There is an irregular field, measuring 100 perches [1] along each side and 50 perches along each end, but 60 perches across the middle. How many acres does it contain?

SOLUTION. The length of the field is 100 perches; the length of the edges is 50 perches, but at the middle it is 60 perches. Join the length of the ends with the middle, which is 160. Take the third part of that, which is 53, and multiply by 100, making 5300. Divide this into 12 equal parts, making 441. Then divide this into 12 equal parts, giving 37. That is the number of acres in the field.

NOTES. 1. The perch (= rod = pole) is $16\frac{1}{2}$ feet or $5\frac{1}{2}$ yards, which is the 320th part of a mile. The Latin is *pertica*, meaning a staff or pole. We use perch as the nearest equivalent, but Prof. Dilke says the *pertica* was 10 Roman feet. The phrasing of what is being measured varies among the various MSS. For acres the Latin is *aripennos*, more correctly spelled *arepennis*. This was a Gallic half-acre — Prof. Dilke says it was about 0.312 of a modern acre. Alcuin has 144 square perches to his aripennos. We use acre as the nearest English term — the modern acre is 160 square perches, but this was not standardized until long after Alcuin and local variations persist to this day.

It will be noticed that Alcuin is ignoring fractions in an inconsistent way. Whereas $\frac{160}{3} = 53.33\ldots$ is reasonably rounded to 53, $\frac{5300}{12} = 441.66\ldots$ is truncated to 441, while $\frac{441}{12} = 36.75$ is not truncated, but rounded to 37. The actual value is $16000/(3 \cdot 12 \cdot 12)$ or $37.037\ldots$, so his overall answer is pretty accurate.

The method of averaging the 50s and the 60 to get $\frac{160}{3}$ is very dubious. If the field is a double trapezium, then the average width is actually 55. Also, the length is given as measured along a side, which may not be the true length of the field. Folkerts & Gericke give a diagram leading to $\frac{160}{3}$ and say Heron gives a similar problem. All in all, this and the following problems show how the Greek knowledge of geometry had been lost to the Roman and medieval worlds.

Folkerts [1] gives an alternative solution which essentially replaces the first sentence by: "Join the two lengths, making 200. Take the half of 200, which is 100." This probably arose from the common Roman and Egyptian "Edfu" approximation for the area of a quadrilateral with sides a, b, c, d, as $\frac{a+c}{2} \times \frac{b+d}{2}$, i.e. the average "length" times the average "width", but Alcuin's calculation of $\frac{160}{3}$ does not fit into this scheme.

23. A four-sided field — *De campo quadrangulo.*

A four-sided field measures 30 perches down one side and 32 down the other; it is 34 perches across the top and 32 across the bottom. How many acres are included in the field?

SOLUTION. The sum of the lengths of the sides is 62 perches, half of which is 31; and the sum of the widths is 66, half of which is 33. 31 times this is 1020 [!]. Divide this into 12 parts, as above, and there are 85; and again divide 85 by 12, and this gives 7. Therefore there are 7 acres here.

NOTES. Here the Roman–Egyptian Edfu area formula, mentioned in the previous note, is clearly being used. However, since $34 + 30 = 32 + 32$, it is possible for such a field to enclose no area at all. The maximum area is obtained when the quadrilateral can be inscribed in a circle, i.e. is a cyclic quadrilateral, when the Archimedes–Heron–Brahmagupta formula can be applied: $A^2 = (s-a)(s-b)(s-c)(s-d)$, with s being the semiperimeter: $s = (a+b+c+d)/2$. In our case, we get $A = 1021.9980\ldots$, so the Roman method gives a good estimate of this maximum area.

24. A triangular field — *De campo triangulo.*

A triangular field is 30 perches along two sides and 18 perches along the bottom. How many acres must be enclosed?

SOLUTION. Join the two long sides of the field and make 60. Half of 60 is 30; because the bottom is 18, take half of this, which is 9; 9 times 30 is 270; divide 270 by 12, which makes $22\frac{1}{2}$; and again divide by 12, giving 1 acre and $10\frac{1}{2}$ perches.

NOTES. There is some gap in some of the MSS here. The Bede text makes an insertion but gives the result as $\frac{27}{12}$ and gets $2\frac{1}{2}$ acres. Folkerts feels that nothing is really missing and we agree.

Square perches are intended. It is still common to confuse perches and square perches in area measurements.

However, it is clear that the result is wrong as $10\frac{1}{2}$ is the remainder in the division of $22\frac{1}{2}$ by 12, whereas the correct surplus is the remainder in the division of 270 by 144, which is 126 ($= 12 \times 10\frac{1}{2}$). That is, the correct answer is 1 acre and 126 square perches or $1\frac{7}{8}$ acres.

25. A round field — *De campo rotundo.*

A round field is 400 perches in circumference. How many acres must it contain?

SOLUTION 1. The fourth part of the field's circumference of 400 perches is 100; multiplied by itself, this gives 10,000. Divide this in 12 parts; the result is 833, which, when divided into 12 parts, produces 69. This is the number of acres in the field.

NOTES on solution 1. This corresponds to $A = \frac{C}{4}^2$, corresponding to taking $\pi = 4$, i.e. to assuming the field is square! The correct area is $\frac{4}{\pi}$ times Alcuin's answer, or 88.419.... Folkerts [1] gives a second solution, interpreted as follows.

SOLUTION 2. The fourth part of 400 is 100 and the third part of 400 is 133. Half of 100 is 50; half of 133 is 66. 50 times 66 is 3151 [!]. Dividing this into 12 parts gives 280 [!!]. Again we divide 280 into 12 parts getting 24. Then 4 times 24 is 96. Thus there are 96 acres.

NOTES on solution 2. The calculation is based on $A = \frac{C}{6} \times \frac{C}{8} \times 4 = \frac{C^2}{12}$, which assumes $\pi = 3$, which is rather better than the first solution, though the calculations are rather cruder than in other problems. The calculation should have given a result of 92.5.... *GIA* [12] starts with $C = 418$ and uses $\pi = \frac{22}{7}$ to get $D = 133$, then multiples $\frac{D}{2}$ by $\frac{C}{2}$ to get 13,898(?) square perches, which is $96\frac{1}{2}$ acres and $2\frac{1}{2}$ square perches.

26. A dog chasing a hare — *De campo et cursu canis ac fuga leporis.*

There is a field 150 feet long. At one end is a dog, and at the other is a hare. The dog chases when the hare runs. The dog leaps 9 feet at a time, while the hare travels 7 feet. How many feet will be traveled by the pursuing dog and by the fleeing hare, before the hare is seized?

SOLUTION. The length of the field is 150 feet. Half of 150 is 75. The dog goes 9 feet at a time; and 75 times 9 is 675; this is the number of feet the pursuing dog runs before he seizes the hare in his grasping teeth. Because the hare goes 7 feet at a time, take 75 times 7, obtaining 525. This is the number of feet the fleeing hare travels before it is caught.

NOTES. The Alcuin text omits the result 525, but Folkerts and Folkerts & Gericke give it. This is the first known European occurrence of this "Hound and Hare Problem". There are many examples of overtaking problems, often more complicated, in Chapters VI and VII of the *Chiu Chang Suan Ching*, c. 100 BCE [8]. Overtaking problems also occur in the Indian and Arabic literature about the time of Alcuin. These generally involve messengers, but the *Chiu Chang Suan Ching* has one (Chap. VI, no. 14) with a hound and a hare, indeed slightly more complex than Alcuin's version. Folkerts [1] thinks this is perhaps a new variant.

27. A four-sided town — *De civitate quadrangula.*

A four-sided town measures 1100 feet on one side and 1000 feet on the other side, on one edge 600 and on the other edge 600. I want to cover it with the roofs of houses, each of which is to be 40 feet long and 30 feet wide. How many dwellings can I make there?

SOLUTION. The two long sides of the town add up to 2100 feet. Similarly, the two short sides add to 1200. Then half of 1200 is 600 and half of 2100 is 1050. Because each house is 40 feet long and 30 feet wide, take the fortieth part of 1050, which is 26; and the thirtieth part of 600, which is 20. 20 times 26 is 520. That is how many houses can be built.

NOTES. *GIA* has "26 with remainder 10". Assuming the shape of the town is an isosceles trapezium and using the Edfu formula, the town is found to have area equal to 525 house areas, though the exact area is 523.17... house areas. As is usual in these problems, Alcuin has made some rounding in his calculations. More importantly, he takes no account of actually fitting the houses into the town.[h]

28. A triangular town — *De civitate triangula.*

A triangular town measures 100 feet along one side and 100 feet along another side and 90 feet along the front. I want to build houses there, each house being 20 feet long and 10 feet wide. How many houses can there be?

SOLUTION. Adding the two sides makes 200, and half of 200 is 100. The front measures 90 feet, and half of 90 is 45. Because each house is to be 20 feet long and 10 feet wide, take one twentieth of 100, which is 5 and take one tenth of 40 [!], which is 4. Then 5 times 4 is 20. That is the number of houses the town should contain.

NOTES. According to the Edfu formula, the area would be 4500, which is $22\frac{1}{2}$ house areas. The actual area is 4018.62..., which is 20.09... house areas. The conversion of 45 to 40 may be a deliberate attempt to compensate for the inaccuracy of the Roman area formula,

[h] I find that one can only get 516 houses in, but if one turns the houses the other way, one gets 517 in. By having some houses each way, I can get 519 in.

or for the difficulty of fitting rectangular houses into the triangular town.[i]

29. A round town — *De civitate rotunda.*

There is a round town, 8000 feet in circumference. How many houses must it contain, each house being 30 feet long and 20 feet wide?

SOLUTION 1. One circuit of the town is 8000 feet. Divided in the proportion of 3 to 2 gives 4800 and 3200. These must contain the length and the breadth of the houses. So take from each one half and there remains of the larger 2400 and of the smaller 1600. Divide 1600 by 20, and there are 80 twenties; and again the greater, that is 2400, divided into thirties, gives 80. Take 80 times 80 and there are 6400. This is the number of houses which can be built in the town according to the above instructions.

NOTES on solution 1. The actual area of the town is 8488.26... house areas. Using $A = C^2/16$ as in problem 25, the area is 6666.66... house areas, while if we use $A = C^2/12$ as in the second solution of problem 25, the area is 8888.88... house areas. Alcuin assumes the circle contains a 1600 by 2400 rectangle, but such a circle would have circumference 9061.73... and would contain 10,890.85... house areas! There are several simple ways to pack houses all in the same direction, starting along a diameter, depending on which way the houses are facing and whether the first row or the first two rows straddle the diameter.[j] Folkerts gives a second solution, with a gap that hc has filled in.

SOLUTION 2. The ambit of this town is 8000. Then take a quarter of 8000, getting 2000. Further take a third of 8000, getting 2666. Take the half of 2000, which is 1000, and the half of 2666, which is 1333. Now the thirtieth part of 1333 is 44, likewise the twentieth part of 1000 is 50; and 50 times 44 makes 2200. Then form 2200 times 4, which is 8800. This is the total number of houses.

---

[i] I can only get 15 houses into the town. John Hadley can get in 16 if the walls can be dented slightly.

[j] I find that having the long side of 30 parallel to the diameter and the first two rows straddling the diameter gives the most houses — 8307. One can probably get more in.

NOTES on solution 2. This assumes that a quarter of the circle contains a 1333 by 1000 rectangle, i.e. a $C/6$ by $C/8$ rectangle, which is related to the belief that $A = C^2/12$. But a circle containing a 2666 by 2000 rectangle would have circumference 10,470.30... and would contain 14,539.75... house areas!

GIA starts with $C = 8008$, uses $\pi = \frac{22}{7}$ to compute $D = 2548$ and then finds $A = \frac{D}{2} \times \frac{C}{2} = 5{,}101{,}096$ square feet and divides this by 600 to get 8501 houses with 496 square feet left over.

30. A basilica — *De basilica.*

A basilica is 240 feet long and 120 feet wide. It is paved with paving stones 23 inches, that is one foot 11 inches long and 12 inches, i.e. one foot, wide. How many stones are needed?

SOLUTION. It takes 126 paving stones to cover the length of 240 feet, and 120 to cover the width of 120 feet. Multiply 120 by 126 and it makes 15,120. That is the number of paving stones required to pave the basilica.

NOTES. The value of 126 arises by rounding up the value of $12 \times \frac{240}{23}$. Folkerts gives a second solution, which assumes the stones are 15 inches by 8 inches but is otherwise essentially the same as the above solution. *GIA* divides the area of the floor by the area of the paving stone, getting 15,026 with the remainder of 24 square inches.

31. A wine cellar — *De canava.*

There is a wine cellar 100 feet long and 64 feet wide. How many casks will it hold, if each cask is 7 feet long and 4 feet wide across the middle, and there is one path 4 feet wide?

SOLUTION. There are 14 sevens in 100, and 16 fours in 64. But 4 of these 64 feet must be taken by the path which runs the length of the cellar. This leaves 60 feet and there are 15 fours in 60. Take 15 times 14, which is 210. That is the number of casks of the given size which can be contained in this wine cellar.

Folkerts gives a second solution, not in *GIA*, where it is assumed that there is a path of width 3 between each row of casks, including one at each side.

> Solution 2. Six sevens are 42 for six rows of casks. And for the paths, which are 3 feet wide, take 7 times 3, which is 21. Hence add 42 and 21, which makes 63. So there are paths for six rows

of casks. Now take the fourth part of 100, which is 25; that is, in each row there are 25 casks. So for six rows, six 25s make 150. This is the number of casks.

NOTES. Hadley observes that problems 27-29 make no provision for paths or streets, in contrast to the present problem. This may indicate that the problems came from different sources, though all of them are also in *GIA*, so they are probably all extracted from some common source. Comparison with the versions in *GIA* makes it quite clear that Alcuin's versions are indeed much cruder. It seems as though some scribe simply avoided copying down hard numbers and made crude approximations of the numbers in his source.

32. A lord of the manor distributing grain — *De quodam patrefamilias distribuente annonam.*

A gentleman has a household of 20 persons and orders that they be given 20 measures of grain. He directs that each man should receive three measures, each woman two measures, and each child half a measure. How many men, women and children must there be?

SOLUTION. Take one set of three measures; that is, one man is given three measures. Similarly, take 5 twos, making 10; that is, five women receive 10 measures. Now, seven twos make 14; that is, 14 children receive 7 measures. Adding 1 and 5 and 14 gives 20; this is the number in the household. And adding 3 and 10 and 7 makes 20, which is the number of measures. So we have both 20 people and 20 measures.

NOTES. The Latin is *modios*, a dry measure of 16 *sextarii*. A *sextarius* is about a pint, so a *modius* is about a peck or a quarter of a bushel. Another source says that it was nearly two gallons. However, the actual size of the *modius* is not really relevant and we shall use the indefinite "measure" for it. There is a second solution: 4, 0, 16, but zero would not have been an acceptable value to Alcuin. See Chapter 12 for a discussion of this type of problem.

33. Another landlord apportioning grain — *De alio patrefamilias erogante suae familiae annonam.*

A gentleman has a household of 30 persons and orders that they be given 30 measures of grain. He directs that each man should receive three measures, each woman two measures, and each child half a measure. How many men, women and children are there?

SOLUTION. If you take three threes, that makes 9; and five twos are 10; and twenty-two halves is 11. That is, three men receive 9 measures, five women receive 10 and 22 children receive 11 measures. Adding 3 and 5 and 22 gives 30 in the household. And adding 9 and 11 and 10 makes 30 measures. So we have both 30 people and 30 measures.

NOTES. In [1] and [2] they just use *Alia propositio* [Another problem], which was generally used to mean: "A Similar Problem." We have used the title given in the Alcuin text. There are two additional solutions: 6, 0, 24 and 0, 10, 20, if 0 is permitted.

33a. Another similar problem — *Item alia propositio.*

A gentleman has a household of 90 persons and ordered that they be given 90 measures of grain. He directed that each man should receive three measures, each woman two measures, and each child half a measure. How many men, women, and children were there?

SOLUTION. Take six threes, making 18; and twenty twos, making 40; and sixty-four halves, making 32. That is, six men receive 18 measures, and twenty women receive 40 measures, and 64 children receive 32 measures. Adding 6 and 20 and 64 gives 90 in the household. And adding 18 and 40 and 32 makes 90 measures. So we have both 90 people and 90 measures.

NOTES. This is the title given by [1] and [2]; the Bede text gives just *Alia*. The Bede text has these problems in a slightly different order so that Problem 33a precedes Problem 33. It is odd that it was not realized that tripling the solution of Problem 33 gives an answer to 33a. This problem actually has 5 positive solutions and 2 more if zero is permitted. The solutions can be very simply expressed as: $3n$, $30 - 5n$, $60 + 2n$, for $n = 0, 1, \ldots, 6$.

34. Another landowner dividing grain among his household — *Altera de patrefamilias partiente familiae suae annonam.*

A gentleman has a household of 100 persons and proposes to give them 100 measures of grain, so that each man should receive three measures, each woman two measures, and each child half a measure. How many men, women and children are there?

SOLUTION. Eleven threes make 33; and 15 times two makes 30. And seventy-four halves is 37. That is, 11 men receive 33 measures, and

15 women receive 30 measures, and 74 children receive 37 measures. Adding 11 and 15 and 74 gives 100 in the household. And adding 33 and 30 and 37 makes 100 measures. So we have both 100 people and 100 measures.

NOTES. In [1] and [2] they just use *Item alia propositio* [Another similar problem]. We have used the title from the Alcuin text, which begins *Propositio altera de....* This problem has six solutions and another if we allow zero values, given by: $2 + 3n$, $30 - 5n$, $68 + 2n$, for $n = 0, 1, \ldots, 6$. This is the same as problem 32 with 20 replaced by 100, so five times the solution of problem 33 is a solution here, but again this had not been noticed.

We wonder if there is some pattern to the solutions of problems 32, 33, 33a, 34. John Hadley observes that in 32, 33, 34, the number of children is the largest even number less than $\frac{3}{4}$ of the household, but in problem 33a this would lead to 66 children rather than the 64 given. However, it is worth pointing out that elimination shows that the ratio of children to the number in the household must lie in the interval $(\frac{2}{3}, \frac{4}{5})$, for which $\frac{3}{4}$ is a reasonable middle value. The exact midpoint is $\frac{11}{15}$, but using this would give 66 children in problem 33a and 72 children in problem 34.

35. A dying man's will — *De obitu cuiusdam patrisfamilias.*

A dying man left 960 shillings and a pregnant wife. He directed that if a boy was born, he should receive $\frac{9}{12}$ of the estate and the mother should receive $\frac{3}{12}$. If however a daughter was born, she would receive $\frac{7}{12}$ and the mother $\frac{5}{12}$. It happened however that twins were born — a boy and a girl. How much should the mother receive, how much the son, how much the daughter?

SOLUTION. 9 and 3 makes 12, and 12 (troy) ounces make a pound. Similarly, 7 and 5 also make 12; and twice 12 is 24. 24 ounces make two pounds, that is, 40 shillings. Hence divide 960 shillings into 24 parts; the twenty-fourth part is 40. Then take nine parts of 40 shillings; the son receives these nine 40s, which is 18 pounds, making 360 shillings. Now the mother takes three parts compared to the son and five compared to the daughter, and 3 and 5 make 8. Then take eight parts of 40. The mother receives eight 40s, that is, 16 pounds, which is 320 shillings. Then take what remains, which makes seven parts, seven parts of 40, which is 14 pounds or 280 shillings. This is what the daughter receives. Adding 360 and 320 and 280 gives 960 shillings or 48 pounds.

NOTES. The reference to ounces arises because the original ratios are expressed as so many ounces (per pound), i.e. the son was to receive 9 ounces (out of every 12). A simplified translation of the rather contorted solution is given in [2].

36. An old man greeting a boy — *De salutatione cuiusdam senis ad puerum.*

An old man greeted a boy as follows: "May you live long — as long again as you have lived so far, and as long again as your age would be then, and then to three times that age; and let God add one year more, and you will be 100". How old was the boy at the time?

SOLUTION. He was 8 years and three months old at the time. The same again would be 16 years and 6 months; double that makes 33 years, which multiplied by 3 makes 99 years. One added to this makes 100.

37. A man wanting to build a house — *De quodam homine volenti aedificare domum.*

A man wanting a house built contracted with six builders, five of whom were master builders and the sixth an apprentice, to build it for him. He agreed to pay them [a total of] 25 pence a day, the apprentice to get half the rate of a master builder. How much did each receive a day?

SOLUTION. Take 22 pence and divide them in six parts. Then to each of the five masters will be given 4 pence, for five fours are 20. The two which remain, which is half of 4, give to the apprentice. That leaves 3 pence which you distribute as follows: divide each penny into 11 parts. Three elevens are 33. Take 30 of these parts and divide them among the 5 masters. Five sixes are 30, so 6 parts go to each master. Take the three parts remaining, which is half of six, and give them to the apprentice.

38. A man buying a hundred animals — *De quodam emptore in animalibus centum.*

A man wanted to buy 100 assorted animals for 100 shillings. He was willing to pay three shillings for a horse, 1 shilling for an ox and 1 shilling for 24 sheep. How many horses, oxen and sheep did he buy?

SOLUTION. Take twenty-three threes, which makes 69. And take twenty-four twos, which is 48. So there are 23 horses and 69 shillings, 48 sheep and 2 shillings, and 29 oxen and 29 shillings. Add 23 and 48 and 29, and there are 100 animals. And add 69 and 2 and 29 and there are 100 shillings. Thus there are both 100 animals and 100 shillings.

NOTE. There is a second solution: 0, 100, 0, if values of zero are allowed.

39. An oriental merchant — *De quodam emptore in oriente.*

A man in the East wanted to buy 100 assorted animals for 100 shillings. He ordered his servant to pay 5 shillings for a camel, one shilling for an ass, and one shilling for 20 sheep. How many camels, asses and sheep did he buy?

SOLUTION. Five nineteens make 95, that is, 19 camels are bought for 95 shillings, being five nineteens. Add another one, that is, buy one ass for one shilling, making 96 [shillings]. Then four twenties make 80, so for four shillings, buy 80 sheep. Then add 19 and 1 and 80, making 100. So there are 100 animals. And add 95 and 1 and 4, making 100 shillings. So there are both 100 animals and 100 shillings.

NOTES. Again there is a second solution: 0, 100, 0, if zero values are allowed. This problem is numerically the same as the first of the problems given by Abu Kamil, c. 900 [4], but Abu Kamil is buying ducks, hens and sparrows. Abu Kamil indicates that the problem is well known and gives several extended forms with four and five types of animals. The coincidence of the numerical values seems to indicate that the problem had already circulated throughout the Arabic and European worlds, but we have no surviving evidence of this. Alcuin may never have seen a camel. On the other hand, camels are mentioned in the Bible, so he would certainly have known of them. Also, he traveled to Rome and he might well have seen camels there. [Henry I, c. 1100, had a camel in his menagerie at Woodstock.]

40. A man and some sheep grazing on a hillside — *De homine et ovibus in monte pascentibus.*

A man looked at some sheep grazing on the hillside, and said: "I wish I had these, and as many again, plus half of half of that total, plus half of that last amount; then, counting myself, I would take a hundred to my home". How many sheep did he see?

SOLUTION. He saw 36 sheep. The same again makes 72, and half of this is 36 and half again makes 18. Again half of 18 is 9. Now 36 and 36 is 72. Adding 18 makes 90. Add 9 to make 99. Adding the man himself makes 100.

41. A breeding sow and a pigsty — *De sode et scrofa.*

A farmer created a new yard in which he put a breeding sow, which produced a litter of 7 piglets in the center pigsty, which, with their mother, makes eight. They bear litters of 7 piglets each in the first corner of the yard; then all of them bear litters of 7 each in the next corner and so on for all four corners. Finally, they all bear litters of 7 in the center pigsty. How many pigs are there now altogether, including the mothers?

SOLUTION. At the first birth, in the center pigsty, there are 7 piglets and the mother makes eight. Eight eights are 64. That is the number of pigs, including mothers, in the first corner. Then eight sixty-fours are 512. That is the number of pigs and mothers in the second corner. Again, eight times 512 is 4096. That is the number of pigs and mothers in the third corner. If one multiplies this by eight, there are 32,768. That is the number of pigs in the fourth corner. Multiplying this by eight gives 262,144. That is the total number of pigs after the final litters are produced in the center pigsty.

NOTES. Alcuin assumes that all the litters are entirely female! The Alcuin text gives 32,788 and 262,304. Folkerts notes that several different answers appear in the various MSS. Prof. Dilke notes that *sode* is not a standard Latin word, but it occurs in all the texts. The classical word is *suile*.

42. One hundred steps — *De scala habente gradus centum.*

A staircase has 100 steps. On the first step stands a pigeon; on the second two; on the third three; on the fourth 4; on the fifth 5, and so on, on every step up to the hundredth. How many pigeons are there altogether?

SOLUTION. We count them as follows. Take the single one on the first step and add it to the 99 on the ninety-ninth step, making 100. Taking the second with the ninety-eighth likewise gives 100. So, for each step, one of the higher steps combined with one of the lower steps, in this manner, will always give 100 for the two steps. However, the fiftieth step is alone, not having a pair. Similarly, the hundredth remains alone. Join all together and we get 5,050 pigeons.

NOTES. Folkerts & Gericke use a ladder with 100 rungs, which is an alternative meaning of the Latin. The sum of an arithmetic progression was well known to the later Greeks, but does not appear in Euclid. It seems to have been known to the Egyptians and Babylonians. However, Alcuin is the oldest version we know of which presents the problem in this kind of fanciful setting. Later European texts frequently have 100 stones (or apples or nuts) set out at regular intervals and one has to go and pick up the first one, bring it back, then get the second one and bring it back, ... These almost always have 100 stones. John Hadley points out that Alcuin presumably combines the pigeons in the manner he does in order to obtain groups of 100 rather than groups of 101 — multiples of 100 being easier for his readers than multiples of 101.

43. A puzzle of pigs — *De porcis.*

A man has 300 pigs and orders that they have to be killed in 3 days, an odd number each day. (There is a similar puzzle with 30 pigs.) What odd number of pigs, either of 300 or of 30, must be killed each day?

SOLUTION. This is a fable. No one can solve how to kill 300 or thirty pigs in three days, an odd number each day. This fable is posed to confuse children.

NOTES. Cardan [7] observes that Euclid IX, 23 shows that the sum of an odd number of odd numbers is odd.

44. A son's greeting to his father — *De salutatione pueri ad patrem.*

A son greeted his father: "Hello, Father." To which his father replied: "Hello, son. May you live long, as much as you have lived. If you triple that number of years and add one of my years, you will have 100 years." How old was the boy at that time?

SOLUTION. The boy was 16 years and 6 months old. Doubled, this makes 33 years. Tripling gives 99. Adding one of the father's years makes 100.

45. A pigeon — *De columba.*

A pigeon sitting in a tree saw others flying and said to them: "I wish there were others with you — the same number again and a third time. Then with the one of me, there would be 100." How many pigeons were there at first?

SOLUTION. There were 33 pigeons which the first saw flying. The same number again makes 66, and a third time makes 99. Add the one that was sitting and there are 100.

46. A man finding a purse — *De sacculo ab homine invento.*

A man, walking along a road, found a purse containing two talents. Others saw this and said: "Friend, give us a portion of your find." He refused to give any to them. So, they set upon him and took the bag from him and each one took fifty shillings. Seeing that he could not stop them, he reached out his hand and snatched fifty shillings for himself. How many men were there?

SOLUTION. A talent is 75 pounds weight or money. A pound of money is 72 gold shillings. Seventy-five times 72 makes 5400, which number when doubled is 10,800. There are 216 fifties in 10,800. That (i.e. 216) is the number of men who were there.

NOTES. A talent (*talentum*) was a variable weight of metal, usually precious. The solution states the value of a talent in pounds and the value of a pound in (gold) shillings. This is vaguely reminiscent of the "Men Find a Purse Problem," where two men find a purse. The first says: "If I had the purse, I'd have twice what you have." The second says: "If I had the purse, I'd have three times what you have." This is an indeterminate problem present in several forms in Mahāvīrā (850) [14] and al-Karagi (c. 1010) [3].[k] It is somewhat related to the Ass and Mule Problem — see problem 16 and Chapter 11. From Fibonacci onward, it has been a standard problem.

---

[k]Perhaps Alcuin's problem is related to "The Bloom of Thymarides", which appears in Diophantos (c. 250) and Iamblichus (c. 325).

47. A bishop dividing 12 loaves among his clergy. — *De episcopo qui jussit 12 panes in clero dividi.*

A certain bishop ordered that 12 loaves should be divided among his clergy. He ordered that each priest should receive two loaves, each deacon one half and each reader a quarter. There are as many clergy as loaves. How many priests, deacons and readers must there be?

SOLUTION. Five twos are 10, that is, 5 priests receive ten loaves. And one deacon gets half a loaf, and 6 readers get a loaf and a half between them. Take 5 and 1 and 6 together, making 12. Again, take 10 and a half and one and a half, making 12. This is 12 loaves, so there are both 12 men and 12 loaves. So the number of loaves and clerics is the same.

NOTE. There is a second solution: 4, 8, 0, if zero is permitted.

48. A man meeting some scholars — *De homine qui obviavit scolaris.*

A man met some scholars and said to them: "How many are there in your school?" One of them answered: "I do not want to tell you that. Count us twice and multiply by three. Then divide into four parts. If you add me to this fourth part, you will have a hundred." How many did he meet while walking along the road?

SOLUTION. Twice thirty-three is 66. This is the number he met when walking. Twice this number is 132. This when multiplied by three makes 396. A fourth part of this is 99. Adding the boy who answered makes 100.

49. Carpenters — *De carpentariis.*

Seven carpenters made seven wheels each. How many carts were fitted?

SOLUTION. Seven 7s make 49. This is the number of wheels they made. 12 times four makes 40 and 8. From 40 and 8 wheels, 12 carts can be built, with one wheel left over.

NOTE. Folkerts & Gericke [2] use "wheelwrights", although the Latin is not that specific.

50. Wine in flasks — *De vino in vasculis.*

How many *sextarii* are there in a hundred measures of wine, and how many *meri* are there in the same hundred measures?

SOLUTION. One measure contains 40 and 8 *sextarii.* A hundred times 48 is 4800. There are that many *sextarii.* Similarly, a measure contains 288 *meri.* A hundred times 288 makes 28,800. There are that many *meri.*

NOTES. Folkerts only gives the title *Alia propositio* [Another problem]. Folkerts & Gericke [2] and we use the title in the Alcuin text. From problem 8, we know that a *sextarius* is about a pint. The Latin for measures is *metra.* The solution tells us that a measure is 48 *sextarii* or 288 *meri,* so is about 6 gallons. The solution tells us that a *meros* is a sixth of a *sextarius,* so is about a teacup. The Alcuin text gives 289 and 28,900.

51. A father dividing flasks of wine — *De vino in vasculis a quodam patre distributo.*

A dying man left to his four sons four flasks of wine. In the first flask there were 40 measures, in the second 30, in the third 20, and in the fourth 10. Calling his steward, he said: "Divide these four flasks of wine among my four sons so that each of their portions shall be equal, both in wine and in flasks." How is it to be divided so that all receive equally from it?

SOLUTION. In the first flask there are 40 measures, in the second 30, in the third 20, and in the fourth 10. Therefore add 40 and 30 and 20 and 10, making 100. Then divide 100 into four parts. A fourth part of one hundred is 25, and double of this number is fifty. Hence each son's portion is 25 measures, and between two sons 50. In the first there are 40 and in the fourth there are 10 measures. These two together make 50. These you give to two sons. Similarly, add 30 and 20 measures, which are in the second and third flasks, and you get 50, and these you give to two of them, as above, and they have 25 measures each, thus making equal division of wine and flasks among the sons.

NOTES. The Latin for measure is *modia* and a *modius* of 16 *sextarii* was about two gallons. See also problem 32. John Hadley comments that the solution is hardly more enlightening than the problem. It seems that some shifting of wine must be intended. See Chapter 15.

52. De homine patrefamilias — A lord of the manor.

A certain gentleman ordered that 90 measures [1] of grain were to be moved from one of his houses to another, 30 *leucas* [2] away. One camel was to carry the grain in three [3] journeys, carrying 30 measures on each journey. The camel eats one measure for each *leuca*. How many measures will remain?

SOLUTION. On the first journey, the camel carries 30 measures over 20 [4] *leucas* and eats one measure for each *leuca*, that is, he eats 20 measures, leaving 10. On the second journey he likewise carries 30 measures and eats 20 of these, leaving 10. On the third journey, he does likewise: he carries 30 measures, and eats 20 of these leaving 10. Now there remain 30 measures and 10 *leucas* to go. He carries these 30 on his fourth journey to the house, and eats 10 on the way, so there remain just 20 measures of the whole sum.

NOTES. 1. The Latin is modia. From Problem 32, we know a modius was about a peck or a quarter of a bushel.

2. See Problem 1 for the size of a leuca.

3. As can be seen, the camel actually makes four journeys.

4. The Alcuin text has 10 here, which leads to considerable confusion.

The above version is basically that in the Alcuin text. Folkerts gives a second version which is basically that in the Bede text, but it is just a rephrasing of the other version and gives exactly the same information.

The amount eaten on the first two journeys is inconsistent with the amount eaten on the third and fourth journeys unless this camel only eats when he is carrying a load! Alternatively, they could start out with three camels and kill two of them at the 20 *leuca* stage.

The optimum solution is for the camel to make two return trips and a single trip to 10 leucas, so he will have consumed 30 measures and he has 60 measures to carry on. He now makes one return and a single trip of another 15 *leucas*, so he will have consumed another 30 measures, leaving 30 to carry on the last 5 *leucas*, so he reaches home with 25 measures.

The key to finding the optimal solution is to have each stage deliver just enough grain to a depot for the camel to carry it on in one fewer journeys. Using this, one can work out the amount remaining for other possible interpretations of the amount eaten.

This problem is variously known as "Crossing the Desert", "The Jeep Problem" or "The Explorer's Problem". This is by far the earliest appearance of such a problem. Four similar problems appear in an unpublished MS of Luca Pacioli, c1500. [See: A. Agostini; *Il "De viribus quantitatis" di Luca Pacioli*; Periodico di Matematiche (4) 4 (1924) 165–192.] Agostini describes the first of these problems and give the titles of the others. The first has 90 apples to be transported 30 miles, with one apple eaten per mile, so is identical to Alcuin! I have also found an elaborate version in Cardan [**Practica Arithmeticae Generalis**; 1539 = **Opera Omnia**, vol. IV, chap. 66, section 57, pp. 152–153] involving carrying food and material up the Tower of Babel! The Tower is assumed 36 miles high and it seems to require 15625 porters. The next appearances of these problems are in the 20C, but in these later versions, one is usually trying to get as far into the desert as possible.

In the middle of the 20C, the problem became very popular. See: Martin Gardner; **The Second Scientific American Book of Mathematical Puzzles and Diversions**; Simon & Schuster, 1961 [= **More Mathematical Puzzles and Diversions from Scientific American**; Bell, London, 1963 = **More Mathematical Puzzles and Diversions**; Penguin, 1966.], chap. 14, prob. 1, for a discussion and references. Pierre Berloquin; **The Garden of the Sphinx**; Scribner's, 1985, gives five variations of the idea as problems 1, 40, 80, 141, 150. In all cases, he is trying to get a man across a desert, rather than as far as possible into it.

(See Loyd [13] and Pierre Berloquin [6], for modern variations.)

53. An abbot with 12 monks — *De homine patrefamilias monasterii XII monachorum.*

An abbot had 12 monks in his monastery. Calling his steward, he gave him 204 eggs and ordered that he should give equal shares to each monk. Thus he ordered that he should give 85 eggs to the 5 priests, and 68 to the four deacons and 51 to the three readers. How many eggs went to each monk so that none had too many or too few but all received equal portions as described above.

Solution. Take the 12th part of 204. This 12th part is seventeen, so 204 is twelve times 17 or seventeen times 12. Just as eighty-five is five times seventeen, so is sixty-eight four times and fifty-one is three times. Now 5 and 4 and 3 make 12. There are 12 men. Again add 85 and 68 and 51, which is 204. There are 204 eggs. Therefore, to each one comes 17 eggs as the twelfth part.

Notes. The problem seems to have been corrupted. Possibly part of the answer has gotten into the question. The similarity of this with problem 47 makes us think that this may be a corruption of a similar problem. For example, if the abbot wished to divide 204 eggs among his 12 monks so that each priest receives 32, each deacon 8 and each reader 4, then there would be 5 priests, 4 deacons and 3 readers, and this solution is unique. If only the numbers 12, 204, 5, 4, 3 survived and the intermediate text were lost, then a careless or ignorant scribe might reconstruct the problem somewhat like the above.

4.5 Summary and Discussions

Given the telegraphic nature of the problems/solutions/notes, this section will try to give a more connected view of Alcuin's *Propositiones*. We will classify the problems and add to the notes. Folkerts gives another classification based on Tropfke [20].

Geometry problems

The geometry problems form a clear group of 12 problems (problems 9, 10, 21–25, 27–31). They involve only a few basic ideas and can hardly be considered as recreations. All of them deal with areas and the method of solution is shown, although no justifications are provided — indeed most of the calculations are based on incorrect formulae and the numbers are often rather arbitrarily modified during the calculation. In a number of cases, the two manuscript types give different solutions, presumably because the later editor was unhappy with the original solutions, though he does not always improve them. The presentations are fairly prosaic. In the simplest cases, one is dividing a rectangle into smaller rectangles, e.g. cutting a cloth into pieces or paving a floor. Other problems involve finding the area of irregular fields. The inaccurate Egyptian–Roman "Edfu" formula for the area of a quadrilateral: $A = \frac{1}{2}(a + c) \times \frac{1}{2}(b + d)$ is used. For a circle, one solution takes π as 4, but an alternative solution takes it as 3. There is a set of problems of fitting rectangular houses into city walls in the form of a trapezium, a triangle and a circle. These are done with no consideration of the actual fitting problem. The actual maximum number of houses that can be fitted in is a non-trivial problem. To see this, try to find out how many 20 by 30 houses can be placed in a circle of circumference 8000 (problem 29). One solution has 6400 houses, the other solution has 8800! From the general ignorance displayed here, we must conclude that Alcuin only had access to Roman material on geometry. Not even the area of the triangle is found correctly! Problems 21-25 and 27-31 appear in an anonymous 9th century geometry, generally with better answers — e.g. π is taken as $\frac{22}{7}$. The two texts are independent and apparently both derive from a lost common source [2].

Well-known arithmetic problems

There are a number of simple calculations (problems 1, 8, 37, 46, 49, 50), but they are often presented in an amusing, even somewhat cryptic, context. For example, problem 1 has a slug going to have lunch with a leech (or a swallow) who lives a league away, but the slug can only go an inch per day. Since a league is 7500 feet, it takes the slug 246 years and 210 days to get to his lunch! (Problem 8 is sometimes cited as a "cistern" problem, but is actually much simpler.) Another group of problems (2-4, 7, 8, 36, 40, 44, 45, 48) are "God greet you" or "Aha" problems. For example, problem 2 leads to one equation of the form $ax = b$. Even the Egyptians gave solutions of such problems by false position, but here the answer is only presented and checked.

There is one problem (42) about an arithmetic progression and two problems (13 and 41) about geometric progressions. Problem 42 has 100 steps with i pigeons located on the ith step. How many pigeons are there? He adds $1 + 99 = 100 = 2 + 98 = \ldots$ and notes that the 50th step is unpaired, so the total is 5050. Although he does not use the anciently known formula for the sum of an arithmetic progression, it seems clear that he did know it and how to obtain it. For example, in a letter of 796 [5] (p. 141), he explains the 153 fish caught by Simon in John 21.11: "If you add each number from 1 to

17, the total is 153; and if you divide 17 into 10 and 7, 10 stands for the Commandments and 7 for the gifts of the Spirit." Though the process of summing an arithmetic progression was well known, problem 42 is the earliest example seen where the problem has been proposed in this somewhat fanciful manner. The next examples seen are in Luca Pacioli, [1, 2] (problem 73). This has 100 objects placed a yard apart along a line and one must pick up the first and return, then pick up the second and return, ... This form remains common down to the 20th century, with 100 oranges or stones or eggs or apples.

The geometric progressions both reduce to simple powers. The first has $1 + 1 + 2 + 4 + \cdots$, so the sums are just powers of 2 and they are individually computed up to 2^{30}. (This is supposed to be the size of an army!) The second is a Fibonacci-like problem of animal breeding, but there is no delay in maturing — a sow has 7 young (all conveniently female) and next year, all 8 of them each produce 7 young. Hence the total number increases as 8^n, up to 8^6 and these values are individually computed. Problem 7 has the geometric progression: 1, 3, 9, 27 in it, but this does not really enter it.

Variants of well-known arithmetic problems

Problem 16 is a variant of the classical "Ass and Mule" problem which has been attributed to Euclid [11] and certainly appears in Diophantos [9] and *The Greek Anthology* [15]. Let $(a, b; c, d)$ denote the general form of the problem: "If I had a from you, I'd have b times you." "And if I had c from you, I'd have d times you." The version attributed to Euclid is (1, 2; 1, 1). Alcuin gives (2, 1; 2, 2) — but he assumes the second distribution takes place after the first has been carried out, which is the same as problem $(a, b; c - a, d)$. Since he has $c = a$ it makes the problem rather easier than usual. Alcuin is the only example seen in this interpretation and this again makes us think that his knowledge of Greek mathematics was minimal. See Chapter 11.

Problem 35 goes back to early Rome. D. E. Smith [19] discusses the history of this "Testament Problem" or "Posthumous Twins Problem." He says that it grew out of the Roman *Lex Falcidia* of 40 BCE which required that at least $\frac{1}{4}$ of an estate should go to the legal heir. He says that the first appearance of the problem is in

Juventius Celsus, c. 75 (which is unseen), and he gives 22 medieval references. Folkerts & Gericke give a Roman version; they quote Cantor's quotation of Salvianus Julianus, c. 140. See [1] or [2] for a general survey of this problem.

The usual version has a man dying with a pregnant wife. He makes a will providing for 2:1 division between son and mother if a son is born, and 2:1 division between mother and daughter if a daughter is born. The division is usually done as 4:2:1 for son : mother : daughter for the case of twins. Alcuin has ratios 3:1 and 5:7. The Roman rule would give 15:5:7, but Alcuin proceeds by first dividing the estate in half and then dividing the two halves according to the given ratios, resulting in 9:8:7. Applied to the usual case, Alcuin's rule would give 2:3:1. The fact that Alcuin's version is quite different from Roman versions again seems to indicate that the original materials were not known. The only other place where Alcuin's method of halving the estate appears is in a Byzantine MS of c. 1305 [22], so perhaps Alcuin learned his version from Constantinople in some way. From about 1400, there were versions with triplets, quadruplets, even triplets with a boy, a girl and a hermaphrodite! Folkerts [1] (p. 33) remarks that this "astonishing" solution shows that the author did not understand Roman legal practice — perhaps the author had never seen the earlier works? Various writers said that the will would be invalid and a 20th century author said that the lawyers would certainly get all the estate!

Problems new to Europe

Two of the *Propositiones* are problems which are known from the Orient but were not known previously in Europe.

Problem 26 is a "Hound and Hare" problem. Basically this is an overtaking problem where the hare has a head-start of some distance, rather than some time as usually occurs. Alcuin's version is very simple — the dog goes $\frac{9}{7}$ the speed of the hare, which has a headstart of 150 feet. Numerous overtaking problems already occur in Chiu Chang Suan Ching of c. 100 BCE [8] and one of them has a hound chasing a hare. Numerous equivalent problems occur in Indian sources.

Problems 5, 32, 33, 33a, 34, 38, 39, 47 (and possibly 53) are examples of the "Hundred Fowls" problem, which originates with

Zhang Qiujian's conclusion of his Suan Jing [Mathematical Manual] in c. 475 [23].

> One rooster is worth five copper cash; one hen is worth three copper cash; three young chicks are worth one copper cash. A man buys 100 fowls with 100 cash; how many roosters, hens and chicks does he buy?

By the 9th century, this problem was well known throughout the literate world. See Chapter 12.

It is notable that Alcuin's problem 39 refers to a man in the Orient buying camels, asses, and sheep. The reference to camels here and in problem 52 might suggest that Alcuin had some Oriental sources. But camels are mentioned in the Bible and Alcuin visited Rome, where camels would have been better known than in York or Aachen — indeed they might even have been some in Rome. So the usage of camels gives us no real clues as to Alcuin's sources. Further, the numbers occurring in problem 39 are identical to Abu Kamil's 9th century version. This is a much more definite indication of some relationship and one might be able to deduce a great deal from this — e.g. that Alcuin must come after Abu Kamil, but it seems clear that Abu Kamil is writing some time after the problem came to the Arabs and so both he and Alcuin may be drawing on some earlier source(s) now unknown.

Completely new problems

The most famous of the new problems here are the three types of river crossing problems (17–20). They are popular in Europe after Alcuin, but no non-European examples have been found. In recent years, versions have been reported from Africa, but no old non-European examples are known. Prof. Grötschel has observed that these are among the earliest combinatorial problems known. This seems essentially correct. The only earlier types of combinatorial problems seem to be mazes — but mazes tend to be unicursal and hence not very interesting until the Renaissance — and magic squares — which are more numerical than combinatorial. (Alcuin's problem 52 is also combinatorial, but also more numerical than combinatorial.) These river crossing problems were sufficiently popular to generate a verse mnemonic; see Chapter 5, problem 6. The problem has had numerous variations, the simplest being to eliminate the marriages and have guards and prisoners, or missionaries and cannibals, where the bad guys must never be allowed to outnumber the good guys, but still keeping to three of each; see Chapter 14.

Problem 6 is the earliest known example of the "Appleseller's problem." Alcuin's version is slightly complicated. It was simplified over the next centuries; see Pacioli's problem 66 in Chapter 6. Such incoherent accounting comes in several other recreational problems such as the "Missing dollar."

Problems 11, 11a and 11b are the earliest examples of what I call "Strange families." The solutions of these are not given in the Alcuin text, but Folkerts gives solutions from other MSS. The "solutions" are pretty vague and nonsensical, leading to statements like "I'm my own grandfather." It has been popular ever since Alcuin.[1] Strange families have also been popular as riddles. There have also been situations where a father and a son married sisters. Are such problems known from Arabic or Roman times?

Problem 12 is the earliest example of a "Barrel Sharing" problem. They are popular in Europe after Alcuin, but no non-European examples have been found yet. How many solutions does such a problem have when there are n barrels of each type? Surprisingly, this does not seem to have been considered before and it is addressed in Chapter 15. Problem 51 is related to problem 12, but it seems to have lost something and the solution involves pouring between barrels, so it is not really of the same form as problem 12.

Problems 14 and 15 are trick problems, apparently new. Problem 14 is hardly mathematical and may well exist in medieval riddle books, but 15 is related to questions like "How many cuts are needed to cut a log into 10 pieces?", etc. Are earlier examples known? Such tricks could have been made by any numerate culture.

Problem 43 is the earliest example seen of another type of trick problem — "three odds make an even." My collection of examples of this kind of problem are all from the 17th century onward. Some early 20th century examples would solve the above as 1, 1, 298, since 298 is "a very odd number of pigs to kill in one day." However, this depends

[1] I have five references to supposedly actual occurrences of the situation.

on a verbal pun which may exist only in English.[m] A different form of the problem involves putting an odd number (e.g. 3) of pigs into an even number (e.g. 2) of pens, with an odd number in each pen. This is solved by having concentric pens (e.g. two pens with 1 in the inner pen and 2 in the space between, so the outer pen actually encloses 3 pigs!). This can be adapted to have 300 pigs in three pens, an odd number in each, by having a large pen containing 297 pigs and two separate concentric pens containing 1 and 2 (totaling 3) pigs. Was some sort of answer like this intended but Alcuin did not know of it?

Problem 52 is by far the earliest example of "Crossing a desert."

Depending on how you count, there are 14–16 types of problems with some novelty, comprising 24–27 problems of the 56 problems in the *Propositiones*. This is an extraordinary proportion — the typical puzzle collection rarely exceeds a level of 20% novelty. The proportion would be even higher if we remove the 12 geometry problems copied from some other source.

Missing problems

There are a number of problems which might have been expected to occur and whose absence may be significant. There are no right triangles anywhere. No sliding spears nor broken trees. This seems to reinforce the idea that the Greek tradition was unknown. Similarly, the lack of any real cistern problem reinforces the same idea. The Loculus of Archimedes was known in 6th century Europe, but seems to have been lost by Alcuin's time. The Chinese Remainder Theorem, although contemporary with the Hundred Fowls, did not migrate as rapidly to the West. Similarly, the Chessboard problem and Magic Squares had not yet got to the West. The Josephus problem is thought to have originated in 8th century Ireland, but the earliest known MS is 9th century and so perhaps it is a bit after Alcuin's time.

Summary

The *Propositiones* is a clear illustration of the Carolingian dawn. Some problems show the loss of the Greek heritage, but a few show

[m]The editor of the *Mathematical Gazette* says that the German word *ungerade* can just about bear both meanings.

some knowledge of it. On the other hand, the number and variety of new (or new to Europe) problems and the general idea of collecting puzzle problems shows that a new spirit is stirring, one which will assimilate new material and revive old material in the following centuries, leading to Fibonacci, Pacioli, and the development of modern mathematics in Europe.

Although few, if any, of the problems can be due to Alcuin himself, he (or whoever did the collecting) has been a remarkable and observant collector. Despite his ignorance of some Greek problems, he must have had access to some Roman and Greek material (it is unclear if Alcuin read Greek, but there would have been some Greek readers about). More importantly, he must have had access to material deriving from the Arabic world, possibly through Byzantium. Despite Alcuin's comment that Einhard could look up the answers in a book, we know of no appropriate books and suspect that many of the problems had not been written down before. Although Alcuin (or whoever) probably did not create any of these problems, the collection of them was an act of great scholarship which established recreational mathematics in Europe.

Bibliography

[1] ALCUIN-FOLKERTS.
[2] ALCUIN-FOLKERTS-GERICKE.
[3] AL-KARAGI.
[4] ASLAM.
[5] S. Allott. *Alcuin of York c. A.D. 732 to 804 — His Life and Letters.* William Sessions, 1974.
[6] P. Berloquin. *Le jardin du Sphinx.* Dunod, 1961. (Translated by Charles Scribner as *The Garden of the Sphinx.* Scribner's, 1985.)
[7] CARDAN II. Chap. 66, section 57, pp. 152–153.
[8] CHIU.
[9] DIOPHANTOS. Book I, Nos. 15, 18, 19, pp. 134–136 in Heath.
[10] Einhard and Notker the Stammerer. *Two Lives of Charlemagne.* Penguin, 1969. (Lewis Thorpe, translator, combines *The Life of Charlemagne* and *Charlemagne.*)
[11] Euclid. c. 325 BCE. in *Euclidis Opera Omnia*, vol VIII. J. L. Heiberg & H. Menge, Teubner, 1916, pp. 286–287.

[12] *Geometria incerti auctoris.* in N. Bubnov, ed., *Gerberti postea Silvestri II papae opera mathematica*, Berlin, 1899; (also Hildesheim, 1963) Appendix IV. (Cited by [5].)

[13] LOYD. The four elopements, pp. 266 & 375. [This material appeared in *Our Puzzle Magazine*, 1, 4 (April 1908).]

[14] MAHĀVĪRĀ(CHĀRYA).

[15] METRODORUS.

[16] T. H. O'Beirne. *Puzzles & Paradoxes.* Oxford University Press, 1965. (also Dover in 1984 and 2017.)

[17] W. F. Richardson. *Numbering and Measuring in the Classical World.* Bristol Classical Press/Duckworth, 1992.

[18] David Singmaster. Triangles with integer sides and sharing barrels. *College Math. J.* 21:4 (Sep 1990) 278–285.

[19] D. E. Smith. *History of Mathematics*, vol. II. pp. 544–546.

[20] JOHANNES TROPFKE, REVISED BY KURT VOGEL, KARIN REICH AND HELMUTH GERICKE. *Geschichte der Elementarmathematik.* 4TH ED., VOL. 1: *Arithmetik und Algebra.* DE GRUYTER, 1980.

[21] LYNN WHITE JR. *Medieval Technology and Social Change.* OXFORD UNIVERSITY PRESS, 1962.

[22] GREEK MS, C. 1305, CODEX PAR. SUPPL. GR. 387, FF. 118V-140V. TRANSCRIBED, TRANSLATED AND ANNOTATED BY KURT VOGEL AS: EIN BYZANTINISCHES RECHENBUCH DES FRÜHEN 14. JAHRHUNDERTS; WIENER BYZANTINISTISCHE STUDIEN, BAND VI; HERMANN BÖHLAUS NACHF., WIEN, 1968. PROBLEM 91, PP. 110–111.

[23] ZHANG QIUJIAN. *Zhang Qiujian Suan Jing* [ZHANG QIUJIAN'S MATHEMATICAL MANUAL]. 468. CHAP. 3, LAST PROBLEM.[n]

[n]There does not seem to be any translation of this book available. The problem has been translated in several works on Chinese mathematics — e.g. Li & Du; Libbrecht; Mikami; Needham.

Chapter 5

The Problems of Abbot Albert

The *Annales Stadenses* were compiled by Abbot Albert of the convent of The Blessed Virgin Mary in Stade. In the midst of these *Annales*, in the year 1152, were inserted a few pages containing 13 recreational problems. Some are the earliest known examples. The other problems are often among the earliest examples known or known in Europe. Consequently, it seems that many people would like to see a translation of this material.

Stade is a small city on the west side of the Elbe estuary a bit downriver from Hamburg. The date of compilation of the *Annales* is uncertain, but it seems to be 1240 (with additions in about 1260). The *Annales* were edited in 1859 [2].

The recreational problems occur on pp. 332–335. The problems are numbered for ease of reference. They are clearly compiled from other sources. Lappenberg [2] attributes the problems to Bede, but that collection is now attributed to Alcuin; see Chapter 4. However, some of the problems do not occur in Alcuin, but occur usually in other earlier collections, such as *The Greek Anthology* problems attributed to Metrodorus [6] — although this text was apparently not known at the time of Albert, its contents reflect the problems known to the late Greeks. Most remarkably, Albert is our earliest known example of the popular jugs problem (problem no. 4).

This chapter is my translation so these problems will be better known.[a] This is a fairly literal translation expressing the sense of

[a]I have used the standard *A Latin Dictionary* of Lewis and Short, but there are many words I did not find there. *The Revised Medieval Latin Word-List* of Latham and personal correspondence are also used.

the original, although the original is sometimes not very clear. Problem 9 is too corrupted to make sense of it. Any corrections to this translation would be appreciated. The Latin *denarius* and *solidus* are translated as penny and shilling. Material that is unclear is indicated with a dagger sign (†).

The list of contents, on p. 279, contains text which is translated as:

3 – Diverse items inserted in the book by Albert in the year 1152.

(a) A list of Roman and German Emperors, additionally of Archbishops of Bremen, signs of the Zodiac, coats of arms, written in the pages of two Guelph codexes.
(b) A notice of Hildegard, abbess of the Sisters of Mount Saint Robert near Bingen, with excerpts from her letters and writings from divine works, therefore called maxims, which have hitherto not been edited.
(c) Narrative of two literate youths Tirri and Firri, proposing arithmetical problems to each other. Such questions seem to have come from the city of Cologne. Similar questions were used between teachers and students in the Middle Ages: but the same can be found in a German poem ascribed to Wolfram von Eschenbach [a noted Meistersinger], where Tyrol, king of Scotland [this may mean Ireland], proposes enigmas to his son Fridebrand in order to instruct him. Further, it is easily seen, by comparing with Albert's problems, that the source of all these is Bede's [now generally ascribed to Alcuin] arithmetic problems if one is not averse to consulting such a book of puzzling questions.
(d) The same two youths discuss itineraries, from Stade to Rome, via Belgium and France, returning via Switzerland or Tyrol.
(e) Further, about a voyage to Jerusalem.
(f) And then about the topography of the Holy Land.

The problems start on p. 332. What follows is our new translation of these. We have given some historical remarks after each problem, in particular the related problems in Alcuin [3] and Metrodorus [6]. Most of these topics are treated at length by Tropfke [9] and Ahrens [1].

So the Virgin Hildegard died in the time of Pope Adrian, when Emperor Friderico reigned. Once upon a time two literate, studious† and curious youths were sitting on Christmas Eve, posing problems between them. One is called Firri, the other Tirri[b].

1. Firri said to his friend Tirri: "If he lives as much as he lived, and as much again and half of that and half of half, he would have one hundred years. How many years has he? Twenty six, said Tirri, and two parts of a year; computed and found to be so. And Tirri continued: Tell me, what rule serves to answer this kind of question? Firri said: I take a number, having regard to the proposition, it is halved and quartered, so evidently four, and so many times I multiply the number by 100, and it makes 400. Then I adjoin everything that the proposition previously stated to the value four. Three times four is twelve, and half of that is two and half of half is one, and they make 15. Next divide the multiplied value of four hundred by 15 and that which you is the number of years lived, and that which is over, so many parts of fifteen, so much he lived of one year. And so one operates in all similar cases."

NOTES. A "part" is a third, so the answer is 26 2/3 years. The solution says that if we start with 4 and carry out the problem, we get 15, so to get 100, we must start with $\frac{4}{15} \times 100$. This is the "method of false position." This generic type of false position problem is also known as an "aha" or "heap" problem, from the Egyptian word used for the unknown. In the Middle Ages, a common form is known as "God greet you" problems because they started with that salutation. See [3] (problems 2–4, 36, 40, 44, 45, 48), [6] (problems 1–4, 50, 116–127, 137–138) and [9] (pp. 573–576).

2. Tirri said to his friend:

> "Six pence are divided among three, in whatever way you want; to tell you, how much each has." And Firri divided. To Volrado he gave one, Otto two, Galterus three, unbeknownst to his friend. Tirri said to double the number of Volrado, nine-tuple that of Otto and ten-tuple that of Galterus. "All this," he said, "produce in one sum and tell it to me." They added and gave him 50. And Tirri said: "Volrado has 1, Otto 2, Galterus 3. Firri said: Give the rule for this.—I will give it, said Tirri.

[b]It is given as Tyrri here but usually it is Tirri.

Six pence having been divided among three, I give one double, another nine times and the third ten times. The number of these added and received by me, I consider how much it differs from 60. I retain the difference, I set it aside, and in the difference I have as many units as I attribute to the one whose amount was nine-tupled; and as many multiples of eight [the Latin is *octonaris* which may denote a coin or some other unit of eight parts] as I attribute to the one who was doubled; and having two parts, I am not ignorant of the third. For example, Volrado who had one, was doubled, hence had two; Otto, who had two, was nine-tupled, hence had 18; Galterus, who had three, was ten-tupled, hence had 30. These were added and 50 was reported. Then considering the difference between this number and sixty, there resulted ten. There is one multiple of eight in ten, and two units. Hence I attribute one to Volrado, whose amount was doubled; two to Otto, who was nine-tupled; and knowing two parts, I know the third. This is the rule of this problem. No multiple of eight shows that the one who was doubled has nothing. No units shows that the one who was nine-tupled has nothing.

Here are all examples.

51.	9.	*One multiple of eight, and one unit.*	1.	1.	4.
50.	10.	*One multiple of eight, and two units.*	1.	2.	3.
49.	11.	*One multiple of eight, and three units.*	1.	3.	2.
48.	12.	*One multiple of eight, and four units.*	1.	4.	1.
47.	13.	*One multiple of eight, and five units.*	1.	5.	
43.	17.	*Two multiples of eight, and one unit.*	2.	1.	3.
42.	18.	*Two multiples of eight, and two units.*	2.	2.	2.
41.	19.	*Two multiples of eight, and three units.*	2.	3.	1.
40.	20.	*Two multiples of eight, and four units.*	2.	4.	
35.	25.	*Three multiples of eight, and one unit.*	3.	1.	2.
34.	26.	*Three multiples of eight, and two units.*	3.	2.	1.
33.	27.	*Three multiples of eight, and three units.*	3.	3.	
27.	33.	*Four multiples of eight, and one unit.*	4.	1.	1.
26.	34.	*Four multiples of eight, and two units.*	4.	2.	
19.	41.	*Five multiples of eight, and one unit.*	5.	1.	
52.	8.	*One multiple of eight, and no units.*	1.		5.
44.	16.	*Two multiples of eight, and no units.*	2.		4.
36.	24.	*Three multiples of eight, and no units.*	3.		3.
28.	32.	*Four multiples of eight, and no units.*	4.		2.
20.	40.	*Five multiples of eight, and no units.*	5.		1.
12.	48.	*Six multiples of eight, and no units.*	6.		

59.	1.	*No multiples of eight, and one unit.*	1.	5.
58.	2.	*No multiples of eight, and two units.*	2.	4.
57.	3.	*No multiples of eight, and three units.*	3.	3.
56.	4.	*No multiples of eight, and four units.*	4.	2.
55.	5.	*No multiples of eight, and five units.*	5.	1.
54.	6.	*No multiples of eight, and six units.*	6.	
60.	nothing.	*No multiples of eight, and no units.*		6.

NOTES. If the amounts initially given to the three people are a, b, c, then they wind up with $2a, 9b, 10c$, which totals to $P = 2a + 9b + 10c$. However, $a + b + c = 6$ and substituting $c = 6 - a - b$ gives $P = 60 - 8a - b$, so the difference $60 - P$ is $8a + b$.

This idea is usually applied to divining a permutation of three things, so that a, b, c are usually taken as a permutation of $1, 2, 3$, but here that is not needed. According to Tropfke [9] (pp. 648–651), the earliest known version of this divination is c. 1075 [8] (p. 109, part IV, no. 17), where $a + b + c = 18$ and the multipliers are 2, 17, 18. The only other example before Albert is Fibonacci [5] (pp. 307–309), who discusses the problem in general, then gives the same example as Albert (without the table) and then does a permutation of 2, 3, 4 and further problems. When a, b, c are a permutation of 1, 2, 3, only six cases can occur and later authors provided mnemonics for remembering which permutation produced which value of P.

3. Firri said further: "There were three brothers at Cologne, who had three casks of wine. The first cask contained 1 bucket, the second 2, the third 3, the fourth 4, the fifth 5, the sixth 6, the seventh 7, the eighth 8, the ninth 9. Divide this wine equally among these three, without breaking any casks." "I do it," said Tirri, "to the oldest, I give the first, fifth and ninth, and he has 15 buckets To the middle one, I give the third, fourth and eighth, and he likewise has 15. So to the youngest I give the second, sixth and seventh; and thus he also has 15, the wine is divided and the casks are not broken."

NOTES. For "cask" the Latin is *vasa*, which denotes any kind of container or vessel. This is also used in the next problem. Cask seems to convey the right image as to size here, while flask or jug seems more appropriate in the next problem. For "bucket" the Latin is *amam*, a form of *hamam* which denotes a water-bucket, particularly a fire-bucket. Although not clearly specified, the problem wants the casks to be equally divided as well as the wine. The solution gives a distribution which can be represented as: 159, 348, 267. There is a second distribution: 168, 249, 357.

This is the first known appearance of this uncommon variant of the classic problem of sharing barrels which has e.g. ten full barrels, ten half-full barrels and ten empty barrels, and which first appears as Alcuin's problem 12, see Chapter 15. However, Alcuin's problem 51 has four barrels containing 10, 20, 30, 40 measures of wine and he divides them among four sons, which requires some transfer of wine. This is clearly a forerunner of Albert's problem, or perhaps Alcuin misunderstood the problem.

After Albert, the problem occurs in two German manuscripts of the 14th and 15th centuries, but then it seems to disappear — no later examples have turned up. Tropfke [9] (pp. 659-660) mentions this only briefly. Additional thoughts on this appear in Chapter 15.

4. "Good, Firri, and I will propose to you another one, and a very subtle one. My master, planning a banquet, sends me to the next city to bring back wine. I took along a flask which held eight measures. I filled it and there was no more in the tavern. You meet me returning in the street, as you are going to fetch wine. You ask where I came from. And I say: 'I come from the forum, bringing wine for my master.' You ask how much. 'Eight measures,' I say. You say to me: 'And I also am going for wine.' I respond: 'There is no more wine.' Therefore you beg me to divide with you. I ask whether you have flasks. You say you have two flasks, one holding five measures, the other three. I therefore will give you half, i.e. four measures, if you can divide [in half] using these flasks. Divide, or go without wine'." Firri said nothing [unclear part], and did not divide. So Tirri: "I will divide if you cannot. Put out your flasks. Behold," he said, "my eight is full, your larger holds five, your smaller three." And he pours five measures into the larger, filling it, and from that, three into the smaller, filling it. Then he pours the three measures from the smallest flask into the largest, 8 measure, flask. Thus there are 6 measures in the largest flask and two in the middle flask. Now he pours two measures from the middle flask into the smallest, and the middle one becomes empty; then he pours 5 measures into the empty middle flask, filling it, and from this he immediately pours one measure into the smallest flask. "Behold," he asserts, "you have four measures in your larger flask." And from the smallest flask, he emptied three measures into his large flask, of 8 measures, and he had 4. "Firri," he said, "thus the wine can be divided."

NOTES. The "measure" mentioned is a *metretus* which was about 9 gallons. This is the earliest known example of this problem which has become a classic. It is discussed at length in Chapter 15. In that chapter we discuss illustrations of the problem from the Columbia Algorism (no. 123) of c. 1350 and in Paolo dell'Abbaco [7] (problem 66) of c. 1440. Until about 1500, all examples are essentially the same problem. See Tropfke, [9] (p. 659).

5. Firri, responding†, spoke thus: "You [must] take a wolf, a goat and a load of cabbages over a water in a small boat which only holds one [of them], without the wolf killing the got nor the goat eating the cabbage." Thus Tirri; "Iabax [c], this is a story for children. Any child of five years first carries over the goat, and returns to bring the wolf and takes back the goat, then carries over the cabbage, and again goes over for the goat."

NOTES. Alcuin's problems 17–20 are the earliest known examples of these river-crossing problems. Albert's problems 5 and 6 may be only the third and second appearances of the two most popular of these — Ahrens [1] (vol. II, 315–316) cites a 12th century MS version of problem 5. The *Columbia Algorism*, c. 1350 [4], has an illustration.

6. "O Firri, I will challenge you with a similar thing. Three men with their wives wish to cross the Rhine. The boat is small, only large enough to hold two persons. How can they be transferred, so that in any crossing, a man protects his wife from others. Let them be called: Bertold and Berta, Gerard and Greta, Roland and Rosa. How can they cross?" Firri hesitating a little, Tirri said: "Berta and Greta first cross. Greta returns and brings over Rosa. So the three women have crossed. Berta returns to Bertold. Gerard and Roland cross to their wives. Gerard and Greta return. Bertold and Gerard cross. Rosa returns alone., the three men remaining on the further shore. Berta and Greta cross to their men. Roland returns to Rosa and crosses over with her. Thus everyone has properly crossed." Having heard this, Firri cheerfully said: "Do you know someway to remember this?"

[c]This seems to be some sort of exclamation.

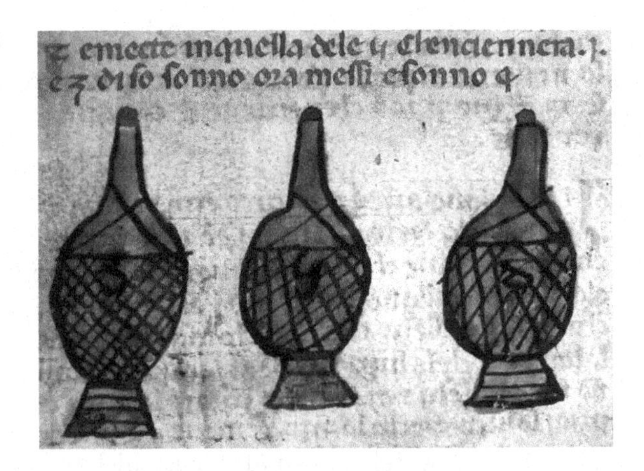

Figure 5.1. Three jugs problem from the Columbia Algorism, c. 1350, no. 122.

Tirri responded: "I know, here are two verses: Binae, sola, duae, mulier, duo, vir mulierque, Bini, sola, duae, solus, vir cum mulier.[d] Firri replied to this: "Explain." And Tirri: "Binae, Berta and Greta; sola, Greta; duae, Greta and Rosa; mulier, Berta; duo, Gerard and Roland; vir mulierque, Gerard and Greta; bini, Bertold and Gerard; sola, Rosa; duae, Berta and Greta; solus, Roland; vir cum mulier, Roland and Rosa."

Notes. Ahrens [1] (vol. I, 6–7) gives another Latin mnemonic verse which was once thought to precede Alcuin, but its date is now uncertain and perhaps 10th century. Again, the earliest illustration is in The Columbia Algorism, c. 1350.

7. "I went," continued Tirri, "to the forum with 30 pence and bought 30 fowls. Geese cost 4 pence; ducks cost two pence; figpeckers cost an obolus. Thus: twenty-four figpeckers for 12 pence, three geese for 12, three ducks for 6."

Notes. For "goose" the Latin is *aucas*. Lewis & Short indicate that this is a form of *avis* and refers to birds in general. But this is close to the Italian *oca* which is goose. For "duck" the Latin is *anetam* which seems like it is related to the Italian *anitra* which is a duck.

[d] An approximate English version is: Women, woman, women, wife, men, man and wife, Men, woman, women, man, man and wife.

Figure 5.2. Columbia Algorism [4], c. 1350, no. 124.

For "figpeckers" the Latin is *fiscedulum* which is close to *ficedula* which is a figpecker. And an *obolus* is half a pence.

Writing the solution in the original order of the birds gives: 3, 3, 12. There is another solution: 0, 10, 20 but solutions with zero values were not accepted at that time.

This is an example of the "Hundred Fowls Problem," which originated in 5th century China and rapidly spread across Asia; see Chapter 12. In Europe, Alcuin's problems 5, 32, 33, 33a, 34, 38, 39, 47 are the first examples. Fibonacci treats the problem extensively; see [9] (pp. 565–569, 572–573, 613–616).

Then there is this unclear passage: Firri said: "It is the vigil of the nativity of our Lord, we must rise before the middle of the night, having slept, we can treat of this in the morning when we part."

8. Tirri said: "To honour this solemnity, three services are offered. A pauper who wished to offer these services, did not have [enough]. So he prayed to God, that he might be given as much as he had in his purse. He received this and offered a penny at the mass at first cockcrow. He prayed again that he would be given as much as he had left. This happened and he again offered a penny. A third time, he obtained as much as he had and giving a penny at the highest mass, nothing was left. Say, Firri, how much he took to the church?" Firri replied: "An obolus, and half an obolus and half of half an obolus."

Tirri demanded: "Give an example." Firri responded to this: "At the first petition, the obolus makes a penny, the half obolus makes a whole and the half of the half makes a half. Thus he gives a penny and retains an obolus and a half. At the second petition, the obolus makes a penny and the half obolus makes a whole. He gives a penny and retains an obolus. At the last petition, this obolus becomes a penny, and the pauper offers this and retains nothing." Tirri responded to this: "It is time to sleep." [*dixisti numerum, eamus dormitum*].

This is an example of the determinate problem whose indeterminate form is the famous Monkey and Coconuts Problem; see Chapter 13. This is the third European version, after the somewhat obscure Abraham: *Liber augmenti et diminutionis* which is a translation from Arabic in the 12th century (or perhaps later) and Fibonacci, who, as usual, treats the problem extensively [5] (pp. 258–267, 278, 313–318 and 329).

9. The following problem is obscure.[e]

Another evening, Firri and Tirri were seated in front of the fireplace and Firri asked Tirri: "What did you eat today?" Firri answered: "Little birds." Then Firri: "I will tell you how many your servant bought. Tell me," said Tirri. Firri replied: "You know what price you paid?" "I know," said Tirri. Firri responded: "For every shilling, I give you a penny. Take my gift and buy a bird. Having bought that, your price leaves a remainder, and as many pence, as can buy one." "I have done it," said Tirri. Firri said: "Your servant bought 13 birds." Replied Tirri: "I understand it well. One obolus given produces 25 birds, a penny 13, two pence 7, three 5, four 4, six 3, twelve two. The result is the number of shillings thought of."

Notes. This seems to be a kind of divination or "think of a number" problem, but there seem to be two unknowns — the number of birds bought and the price. However the price references may simply indicate ratios, i.e. "for every shilling, I give you a penny" may mean "for every twelve, I give you one." The number pairs, x, y, which are given in the answer are related by $(x + 12)/x = y$ or $x = 12/(y - 1)$, where y seems to be the number of birds and x seems to be a price.

[e] I asked the late Prof. Dilke about this and he said that the text is very corrupt and some lines are missing. He sent me a letter with a reconstruction and a translation. This is my translation.

Then $xy = x + 12$ might be the cost of the birds. It seems that in some way, Firri gets Tirri to produce x and then Firri determines y from it.

10. And Firri added: "I will tell you how many pence you have in your purse." Firri said: "I know what I have." And Tirri: "Add twice as much." He did so. Again Tirri: "Take away half." He took it away. And Tirri: "To the half remaining, again add twice as much." He added. Tirri said: "Again take away half." Again he took it away. And Tirri: "Was there an obolus in the first division?" "No," said Firri. Then Tirri: "Was there one in the second?" Firri: "No." "You know what you have left?" asked Tirri. "I know," he responded. And Tirri: "Give Walter nine from the residue." "I give," he said. Again Tirri: "Give nine to Otto." "I cannot," said Firri. Tirri added: "You have four pence in your purse." Firri responded: "Tell how to remember this." Said Tirri: "Think of an amount, add twice to it, and divide, taking away half. Do you have an obolus in the division?"—"I have."—"Make it a pence. Give nine to Walter."—"I cannot." "Then I know that you have three pence. And to thoroughly show you, one penny at the first division gives an obolus, in the second none, and nine cannot be given. Two pence gives an obolus in the second division, in the first none, and 9 cannot be given. Three pence gives an obolus in each division, and 9 cannot be given. Four has an obolus in neither division, and 9 can be given. Five, like one, gives an obolus in the first division, none in the second, but 9 can be given. Six, like two, gives an obolus in the second division, none in the first, and 9 can be given. Seven, like three, in both, and 9 can be given. Eight, like 4, in neither, but 9 can be given twice. Thus one can proceed to infinity, nine having in division like one and 5, ten like two and 6, eleven like three and 7, twelve like four and 8, and so on successively, except that for every multiple of four, one adjoins nine. For example, one can give 9 once from 4 until eight, twice from 8 until 12, thrice from 12 until 16, four times from 16 until 20, and so on further. Other numbers are in between, i.e. between the ascending multiples of four, according to whether there are oboli in the divisions. One can work out how to deal with shillings, marks and pounds in the same way." Firri said: "I consider this in another way. I take a handful of peas, so many as the number, and I use the peas to divide the number into three parts, and discard the excess, which cannot be other than one or two. Divide the [resulting] number in three parts, working with the third part according to the already stated mode of pence, and finding the third part, I know the two remainders, and that which has been discarded, I know also in this way." Tirri said: "And I consider this in another way. Line up three or four or more men and I give pence among them. 'Whoever wants, discard,' I say to them. I give the first one, the second two, the third three, the fourth four and so on successively. This done, I say thus: 'Whoever discarded pence, knows well how much he had. Add the half part of this to this, and if there an obolus in the addition, make if a penny. Again add the half of that which he has in total, and if this has an obolus, make it a penny.' By such addition, and considering the oboli, and the number of nines contained, the number denoted above occurs to you. If one is found, the first discard pence, if two, the second, if three, the third, and so forth."

NOTES. For "Walter" the Latin was *Woltero*.

This is really a simple manipulation in binary arithmetic, but tripling before dividing by two perverts it a bit and the process is obscured by the verbiage required when algebra is not available. Let $n = 4a + 2b + c$, where $b, c = 0$ or 1. The process converts this to $12a + 6b + 3c$, then $6a + 3b + 2c$, then $18a + 9b + 6c$ and finally to $9a + 5b + 3c$. The fact that one takes away the smaller half when the number is odd is not clear in the text but can be deduced from the details. There is an obolus in the first division if and only if $c = 1$; and an obolus at the second division if and only if $b = 1$. The number of nines that can be given is a. The next to last paragraph seems to be an alternate way of doing the problem, but it is too vague to understand!

See [9] (pp. 642–651, especially pp. 643–644), where he discusses this type of problem, but does not mention Albert, who would be the fourth oldest example, after pseudo-Bede, Fibonacci and Johannes Hispalensis. This type of problem may not have been studied before.

11. Firri similarly said: "A certain servant was going from Cologne to Nussiam when he met another, hurrying from Nussiam to Cologne. They saluted one another. After salutations, each told the other of the hospitality of his mother, and each was pleased with this hospitality. One was named Bertold, the other Stephan. Bertold enjoyed the hospitality of Stephen's mother and took her as his wife. Her name was Berta. Stephen enjoyed the hospitality of Bertold's mother and likewise took her to wife. And she was named Osanna. Each of the couples produced a son. What is the relation between them?" Tirri said: "First I draw for you a clear representation, and in this distinctness, I will show you that each is the paternal uncle of the other."

Bertold — s. of Osanna & b. of Galter; Berta — m. of Stephen
Conrad
son of Bertold & paternal uncle of Galter

Osanna — m. of Bertold; Stephen — s. of Berta & b. of Conrad
Galter
son of Stephen & paternal uncle of Conrad

Words of Berta to Galter, or Osanna to Conrad. "Dear boy, your father was borne by me, And you call me grandmother, and I call your brother husband."

NOTES. The Latin *Nussiam* is the modern city of Neuss, about 40km downriver from Cologne, near Düsseldorf. This and the next two problems are examples of what I call "Strange Families." This and problem 12 are essentially identical, except that Latin distinguishes between paternal and maternal uncles. Such problems arise as ancient riddles — an unreferenced source asserts that the Queen of Sheba posed a relationship riddle based on Lot and his daughters to Solomon. The earliest known examples of the present type are in Alcuin. His problem 11 has two men who marry the other's sister. His problem 11a (i.e. it occurs in the "Bede" version of the text) is identical to Albert's problem 11. Alcuin's problem 11b has a father and son marrying a daughter and mother. Tropfke [9] has a brief mention of these on p. 660. [1] (vol. I, 156–162), discusses these problems. (Abbreviations in the figure are for son, brother and mother.)

"Strange Families" riddles appear in *The Exeter Book*; translated and edited by Kevin Crossley-Holland; (*The Exeter Riddle Book*, Folio Society, 1978, Penguin, 1979); revised ed., Penguin, 1993. [Exeter Cathedral Library MS 3501, also known as the *Codex*

Exoniensis, is a 10th century MS.] Problem 43, pp. 47 & 103. Unfortunately, the riddles of the Queen of Sheba are not recorded in the Bible (I Kings 10 and II Chronicles 9).

> Body and soul both have the earth as their mother and sister. Their mother because they are made from dust; their sister because all are made by the same heavenly father.

Problem 46, pp. 50 & 104.

> A man sat sozzled with his two wives,
> his two sons and his two daughters,
> darling sisters, and with their two sons,
> favoured firstborn; the father of that fine
> pair was in there too; and so were
> an uncle and a nephew. Five people
> in all sat under that same roof.

The solution is given in Genesis 19:30–38, which describes Lot and his two daughters who bore sons by him. "The first use of this incestuous story for the purpose of a riddle is attributed to the Queen of Sheba; she tried it on Solomon." Unfortunately, the riddle of the Queen of Sheba are not recorded in the Bible (I Kings 10 and II Chronicles 9). If this is true, it would be by far the earliest known strange families riddles.[f]

12. And Tirri continued: "I give you a similar case for consideration by the following figure. Two soldiers had two daughters. Each of them married the daughter of the other. These pairs produced two sons. If the soldiers are unrelated, and likewise the two daughters, it is a worthy question: what relation exists between the boys."

Florentius f. of Flora	Gertrude d. of Gerard & s. of Jacob

Peter
maternal uncle of Jacob

Flora d. of Florentius & s. of Jacob	Gerard f. of Gertrude

Jacob
maternal uncle of Peter

[f] I have been told that these riddles appear in the Hebrew Targum (annotations) to the Book of Esther, but I have not yet seen an English version of this.

NOTES. Each of these is the maternal uncle of the other, as Gertrude is sister to Jacob and Flora is sister of Peter. (Abbreviations in this figure are for father, daughter and sister.)

13. Firri said further: "My master who never wants to take, not any thing [unclear part], first heard this worthy question. He bought something. He returned and announced the following. Toward [unclear part] me in the street appeared 12 old men. Having saluted them, I continued. Then I further met 12 men of middle age. After a short time, I thirdly met 12 youths, who preceded a decrepit old man, who led with them [unclear part] a quite young girl. The old man saluted me reverently: "You see," he said, "this girl? She is my wife, the twelve youths are our children, the 12 middle aged men whom you met, are my sons, and paternal uncles to the girl, the 12 men are also my sons and all are maternal uncles of the girl. Good Tirri, explain this generation to me." Tirri said: "I do so, and thus:"

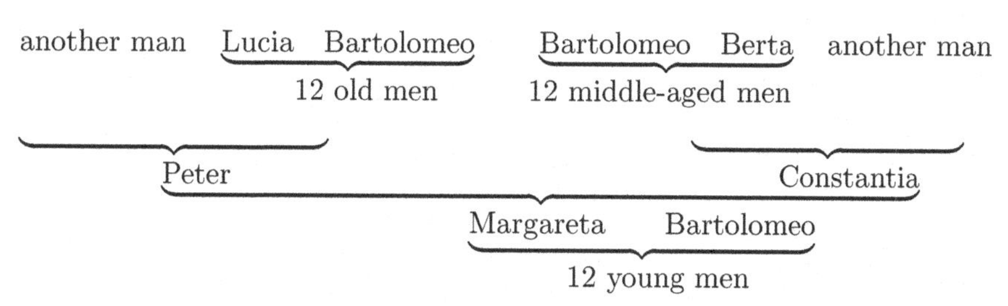

Firri further said: "Good Tirri, I wish to go to Rome, explain the itinerary to me." Replied Tirri: "What way do you want to go?" And he: "Toward the valley of the Moors; but first first to Dacia by horse, so proceeding there from Stade. ...".

NOTES. The diagram contained more text from which we learn: Bartolomeo is the decrepit old soldier, Lucia is the wife of the old man who produced Peter from another man, Berta is the wife of the old man who produced Constantia before marrying the old man, Margaretia is the wife of the old man and is the daughter of Peter and Constantia; Lucia and the old soldier produced the twelve old men who are the paternal uncles of Margareta, Berta and the old man produced the twelve middle-aged men who are the maternal uncles of Margareta, and Margareta and the old man produced the twelve young man.

The valley of the Moors in Latin is Maurianum. Folkerts informs us that this is Maurienne (= Garocella) in Savoy. Dacia is the area of modern Romania and Moldova.

5.1 Summary

Problems 3, 4, 13 are first known examples. One might count problem 12 as new. Problems 5, 6, 11, 12 are second known examples, after Alcuin. Problem 2 is the third known example, after Ṭabarī and Fibonacci. Problem 10 is the fourth known example. Problems 7, 8 are the third known European examples — in both cases Fibonacci is the second known European example. Problem 9 is corrupted. Problem 1 is ancient.

Bibliography

[1] AHRENS.
[2] ABBOT ALBERT.
[3] ALCUIN.
[4] COLUMBIA ALGORISM.
[5] FIBONACCI.
[6] METRODORUS.
[7] DELL'ABBACO-ARITMETICA.
[8] ṬABARĪ .
[9] J. Tropfke, revised by Kurt Vogel, Karin Reich and Helmuth Gericke. *Geschichte der Elementarmathematik.* 4th ed., Vol. 1: Arithmetik und Algebra. De Gruyter, Berlin, 1980.

Chapter 6

Pacioli: The First Book of Mathematical Puzzles

Luca Pacioli (c. 1445–1517) was born and probably died in San Sepolcro (=Borgo), a small city in southeastern Tuscany and he is sometimes called Luca del Borgo.[a] He was the most influential mathematician of his day. Pacioli was a leading expositor of the new theory of perspective, he revolutionized accounting and was the author of the most important mathematical work after Fibonacci.

He wrote several important mathematical treatises. We will focus on the *De Viribus Quantitatis* which was the first volume devoted to recreational mathematics, a book that had the first appearance of many puzzles still well-known today. We give his biography to put these works in context.

He taught in San Sepolcro, Venice, Perugia (first professor of mathematics there), Rome, Zara (on the Dalmatian coast, then part of Venice), Naples, Milan, Florence, Pisa and Bologna. His life was remarkably peripatetic, even for the day. There are indications that his superiors in the Franciscan Order advised against his teaching boys and perhaps this behaviour led to his having to move frequently. He spent some time in Venice, c. 1466–1470, teaching in the house of Antonio de Rompiasi in the Giudecca, and attended lectures of Domenico Bragadino at the Scuola di Rialto. Pacioli stayed with

[a]This is a much expanded version of the text of my part of a joint presentation at Gathering for Gardner 6, Atlanta, Mar 2004. Vanni Bossi's part follows as the next chapter. With thanks to Dario Uri and Bill Kalush.

Alberti in Rome for about a year from 1470. He later became attached to Cardinal Francesco della Rovere and lived at his palace adjacent to the basilica of S. Pietro in Vincoli. In 1475 Pacioli entered the Franciscan Order in San Sepolcro.

Luca Pacioli was the first mathematics lecturer at the University of Perugia, in 1477–1480. Pacioli states in his *Summa* that he was here in 1475–1480 and that in 1475, he lectured to a class of 150. Pacioli spent 1481–1486 teaching at Zara, on the Dalmatian coast and then part of Venice. He was lecturing at the University of Perugia from 1486–1488, and in 1489 Pacioli was in Rome, at the palace of Cardinal Giuliano della Rovere.

In the 1490s Pacioli taught in Naples, possibly completing his *Summa* there. He gave lectures on mathematics in San Sepolcro in 1493. In 1494, Pacioli published the greatest mathematical work since

Sũma de Arithmetica Geometria Proportioni et Proportionalita.

Continentia de tutta lopera.

De numeri e misure in tutti modi occurrenti.
Proportioni e pportiõalita anotitia del. 5° de Euclide e de tutti li altri soi libri.
Chiavi overo evidentie numero. 13. p le q̃tita continue pportiõali del. 6° e. 7° de Euclide extratte.
Tutte le pti dealgorismo: cioe relevare. ptir. multiplicar. sũmare. e sotrare cõ tutte sue pue i sani e rotti. e radici e progressioni.
De la regola mercantesca ditta del. 3. e soi fõdamenti con casi exemplari per c° m° §. G. guadagni: perdite: transportationi: e investite.
Partir. multiplicar. summar. e sotrar de le proportioni e de tutte sorti radici.
De le. 3. regole del catayn ditta positiõe e sua origie.
Evidentie generali over conclusioni n° 66. absolvere ogni caso che per regole ordinarie nõ si podesse.
Tutte sorte binomij e recisi e altre linee irratiõali del decimo de Euclide.
Tutte regole de algebra ditte de la cosa e lor fabriche e fondamenti.
Compagnie i tutti modi. e lor partire.
Socide de bestiami. e lor partire
Fitti: pesciõi: cottimi: livelli: logagioni: egodimenti.
Baratti i tutti modi semplici: composti: e col tempo.
Cambi reali. secchi. fittitij. e diminuti over comuni.
Meriti semplici e a capo danno e altri termini.
Resti. saldi. sconti. de tempo e denari e de recare a un di piu partite
Ori. argẽti. e lloro affinare. e carattare
Molti casi e ragioni straordinarie varie e diverse a tutte occurentie commo nella sequente tavola appare ordinatamente de tutte.
Ordine a saper tener ogni cõto e scripture e del quaderno in vinegia.
Tariffa de tutte usanze e costumi mercanteschi in tutto el mondo.
Pratica e theorica de geometria e de li. 5. corpi regulari e altri dependenti.
E molte altre cose d̃ grandissimi piaceri e frutto cõmo difusamente per la sequente tavola appare.

Ad illustrissimum Principem Gui. Ubaldum Urbini Duce Montis feretri: ac Durantis Comitem. Grecis latinisq; litteris Ornatissimum: et Mathematice discipline cultorem ferventissimum: Fratris Luce de Burgo sancti Sepulchri; Ordinis minor: et sacre Theologie Magistri. In artes arithmetice: et Geometrie. Prefatio.

A quantita Magnanimo Duca: e si nobile et excellẽte cosa che molti phylosophi per questo lhano giudicata ala substantia para: ecõessa coeterna. Peroche hano cognosciuto per verũ modo alcuna cosa in rerũ natura sença lei nõ potere existere. Per la qual cosa de lei lẽdo (cõ laiuto de colui che li nostri sensi reggi) tractarne: nonche per altri prischi e antichi phylosophi nonne sia copiosamente tractato: e in theorica e pratica. Ma per che lor dicti gia ali tempi nostri sonno molto obscuri: e damolti male apresi: e ale pratiche vulgari male applicati: diche i loro operationi molto variano: e con grandi elaboriosi affanni mettano in opera: si de numeri cõmo de misure: unde di lei parlando non intendo se non quãto che ala pratica e operare sia mestiero: mescolandoci secõdo iluoghi oportuni ancora la theorica: ecausa de tale operare: si de numeri cõmo de geometria. Ma prima accio meglio q̃llo che sequita se habia apprehẽdere: essa quantita dividiremo secõdo el nostro proposito: edividendola aciascun suo membro assegnaremo sua propria e vera diffinitione e descriptione. E alora poi sequira quello che Arist. dici in secundo poster. Tũc enim maxime scimus aliquid cum habetur suum quid est etc.

Figure 6.1. First page of the contents of the *Summa* and the title page.

Fibonacci (1202). *Summa de Arithmetica, Geometria, Proportioni et Proportionalità*, Venice, 1494. This is a massive book of 616 large pages. Part II, ff. 68v–73v, problem 1–56, is essentially identical to Piero della Francesca's *Trattato*, ff. 105r–120r. It contained the first printed pictures of Archimedean polyhedra — truncated tetrahedron and cubo-octahedron. He also mentioned the icosidodecahedron and truncated icosahedron.

He came to Venice to publish his *Summa* [1]. This was the first printing of many mathematical concepts, e.g. algebra, double-entry bookkeeping, pictures of some Archimedean polyhedra. He asserted that cubics and quartics cannot be solved by the methods used for quadratics, which inspired the development of algebra. Pacioli spent some time in Venice supervising the publication. See Chapter A for charges of plagiarism of Piero della Francesca [4] against Pacioli.

The *Summa* contained several novel recreational topics. First was the "Problem of Points": If you agree to play until one player has won three times, how do you divide the stakes if you have to stop, say when the score is 2–1? (This later was one of the sources of probability theory.) Second was "Locate a Well Equidistant from the Tops of Three Towers."

He was a professor at Milan in 1496–1499. He was inspired to start his *Divina Proportione* on 9 February 1498 and completed it on 14 December 1498, though it was not published (in an expanded form) until 1509. See Figure 6.3. The period in Milan was the high point of his career, being a leading member of the glittering intellectual court of Lodovico Sforza. He lived at the monastery of San Simpliciano, writing his *Divina Proportione*, and much of *De Viribus Quantitatis* here. He was a good friend of Leonardo da Vinci (1452–1519) who drew the pictures for *Divina Proportione.* (Pacioli seems to have made several sets of models of the polyhedra in his book, though we do not know if Leonardo assisted in making them or used them for his drawings.) Pacioli is our leading witness to Leonardo's work at this time, particularly the "Last Supper" during 1495–1497, and he probably advised on the perspective of the painting. Certainly Pacioli stimulated Leonardo's interest in perspective and it is possible that Leonardo's famous drawing of the proportions of the human body, see Figure 6.2, was inspired by Pacioli's comment on classical architecture; "For in the human body they found the two main figures ..., namely the perfect circle and the square."

When the Sforzas were overthrown by the French invasion in 1499, he and da Vinci left Milan together and came to Florence, originally lodging in the same house. Later he was living in the monastery of Santa Croce, while teaching at the Universities of Florence and Pisa. He spent most of 1500–1507 here. He may have taken some time out to teach at Bologna and perhaps Dürer studied under Pacioli in Bologna in 1506. (Dürer commented in 1506 that he was going to Bologna to see the leading authority on perspective. Dürer certainly studied with de Barbari.)

Pacioli returned to Venice in 1508–1509 to publish his *Divina Proportione* [2] and gave a lecture on the Fifth Book of Euclid at the Church of S. Bartolomeo. The famous portrait of Pacioli by de' Barbari is thought to have been painted in Venice at this time. Erasmus was also in Venice at the time and may have attended Pacioli's lecture — he certainly satirizes Pacioli and his works in *In Praise of Folly*. Pacioli's biographer feels that Pacioli probably stayed at the monastery of San Rocco. Dürer seems to have taken ideas from Piero della Francesca's *De Prospettive Pingendi* which was at Urbino and Pacioli is the most likely person to have shown it to Dürer.

He retired to San Sepolcro as Commissioner (head or warden) of the Franciscan House in 1510, though he went briefly to teach in Perugia in 1510 and in Rome in 1514. Pacioli probably died in San Sepolcro, but there is no gravestone or grave site. There is a street named after him.

Pacioli is the earliest mathematician of whom we have a genuine portrait. This splendid picture is in the Museo Nazionale Capodimonte in Naples, apparently by Jacopo de' Barbari, probably done in Venice, about the time of publication of Pacioli's *Summa* of 1494 (or of *De Divina Proportione* in 1509). The books depicted are his *Summa* and the first printed Euclid of 1482. Dürer came to Venice during his Wanderjahre in 1493, and knew Jacopo de Barbari.

It is claimed that Pacioli is the second figure from the right in Piero della Francesca's "Madonna and Child, with Saints and Angels," done about 1472. The depicted saint is St. Peter Martyr, distinguished by the gash on his head.

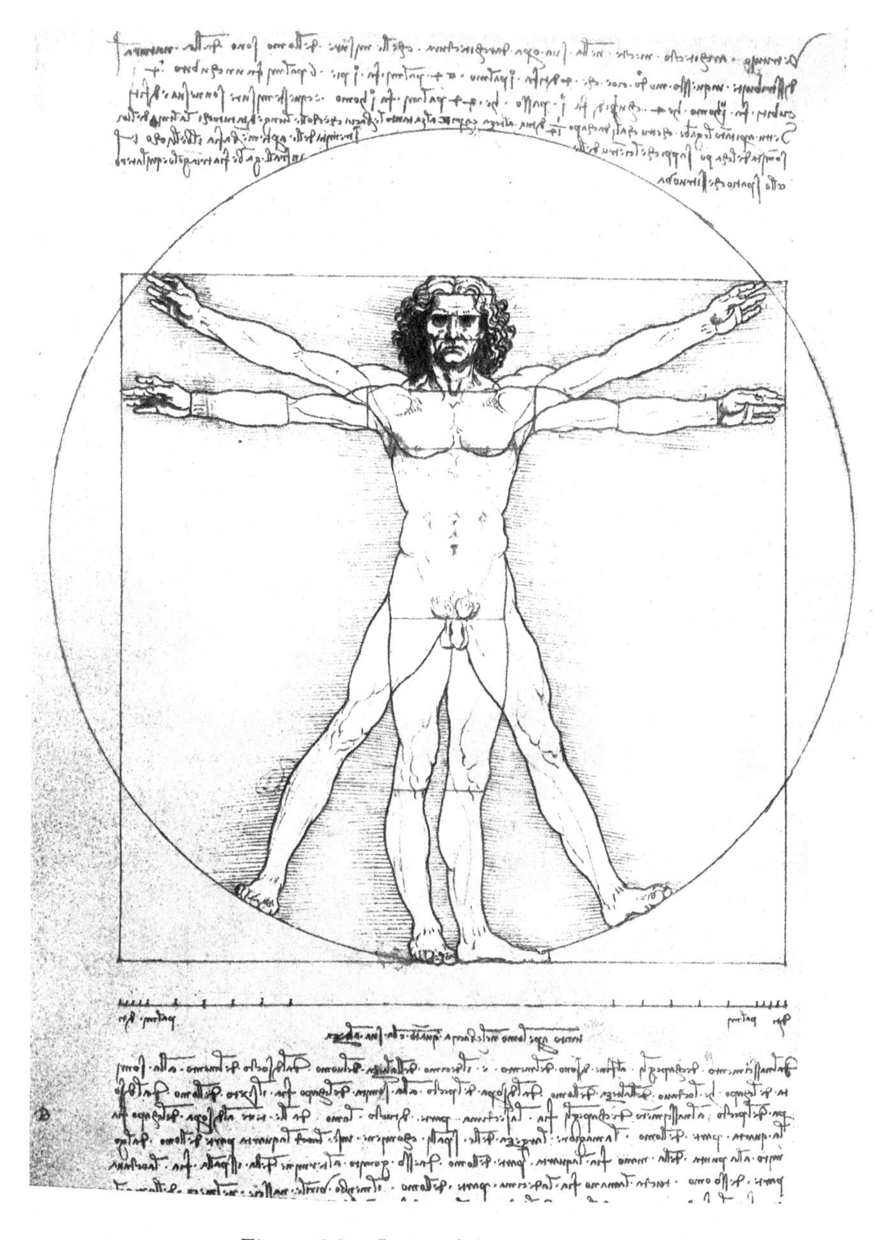

Figure 6.2. Leonardo's geometric man.

6.1 *De Viribus Quantitatis*

We focus on *De Viribus Quantitatis* of c. 1500. This is an Italian MS in Codex 250, Biblioteca Universitaria di Bologna. Pacioli petitioned for a privilege to print this in 1508 and a problem has a date of 1509, but he seems to have been working on the MS since 1496. The title is a bit cryptic, perhaps the best English version is *On the Powers of Numbers*. The great variety of novel problems is emphasized below.

Dario Uri has photographed the entire MS and enhanced the images and put them all on a CD.[b] This includes the whole text, including the indexes with his comments and diagrams of later examples of the puzzles. All the figures from *De Viribus Quantitatis* are from Uri's photos unless specified.

There is a transcription by Maria Garlaschi Peirani, with Preface and editing by Augusto Marinoni; Ente Raccolta Vinciana, Milano, 1997.[c] This text is cited as Peirani. The transcription is not exactly literal in that Peirani has expanded abbreviations and inserted punctuation, etc. Also, Peirani seems to have worked from the microfilm or a poor copy as she sometimes says the manuscript has an incorrect form which she corrects, but Dario Uri's photo clearly shows the MS has the correct form. Peirani uses the problem numbers and names in the MS (these sometimes differ from those in the Contents and Index), but with some amendments. The problem names are given as in the MS, with some of Peirani's amendments. This is the first large work devoted to recreational mathematics. There are three parts. It opens with a Table of Contents, which turns out not to be very dependable.

The actual text opens with some fragments of dedication, possibly leaving room for some fancy initials.

- **Part 1** *Delle forze numerali cioe di Arithmetica* is 81 arithmetical recreations.

---

[b]This has 614 images, including the insides of the covers. This is often more legible than the microfilm version, but the folio numbers are often faint, sometimes illegible. He has put much material up on his website: digilander.libero.it/maior2000/.

[c]Dario Uri says it can be bought from: Libreria Pecorini, Milano.

proportione
Opera a tutti glingegni perſpi
caci e curioſi neceſſaria Oue cia
ſcun ſtudioſo di Philoſophia:
Proſpectiua Pictura Sculptu
ra: Architectura: Muſica: e
altre Mathematice: ſua
uiſſima: ſottile: e ad
mirabile doctrina
conſequira: e de
lectaraſſi: cõ va
rie queſtione
de ſecretiſſi
ma ſcien
tia.

M. Antonio Capella eruditiſſ. recenſente:
A. Paganius Paganinus Characteri
bus elegantiſſimis accuratiſſi
me imprimebat.

Figure 6.3. Title pages of *De Divina Proportione* MS and the printed version.

- **Part 2** *Della virtù et forza geometrica* (on geometric virtue and power) contains 134 geometrical and topological problems.
- **Part 3** contains several hundred proverbs, poems, riddles and tricks (i.e. physical recreations, conjuring, etc.), in several sections: Very Useful Moral Items like Proverbs (23 rhyming couplets); Lament of a Lover for a Maid (27 rhyming couplets, based on the letters of the alphabet and some extra couplets); Very Useful Mercantile Items and Proverbs (83 items); On Literary Problems and Enigmas (about 80 items); Common Problems to Exercise the Ingenuity and for Relaxation (222 items).

Part 1 is described by Agostini [5] whose descriptions are sometimes quite brief — unless one knows the problem already, it is often difficult to figure out what is intended. Further, he sometimes gives only one case from Pacioli, while Pacioli does the general situation

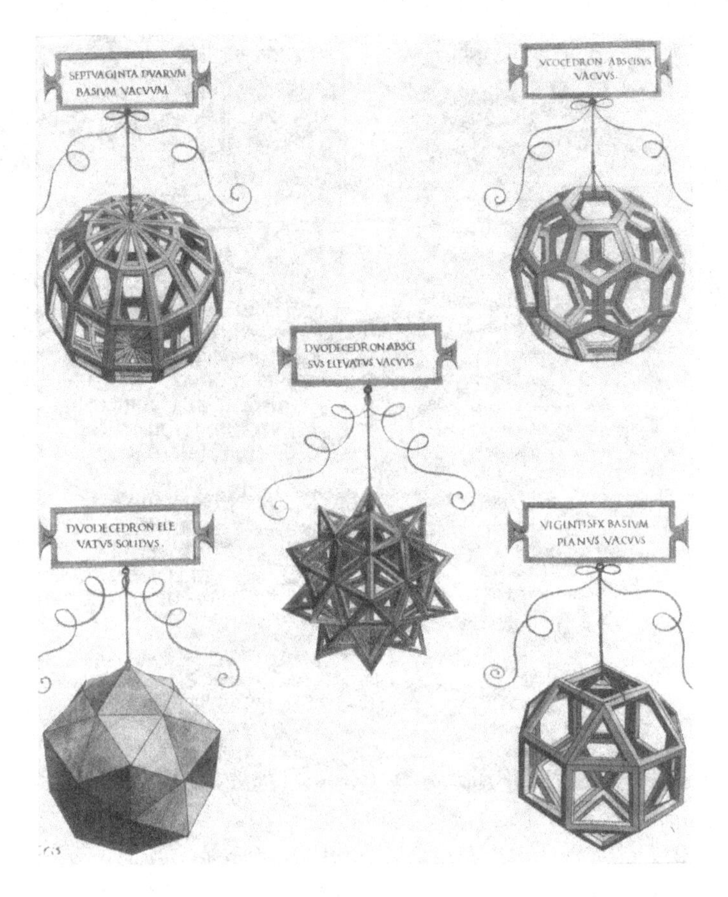

Figure 6.4. Several pictures from the MS, from a poster.

and all the cases. There are 81 problems in Part 1, but the Index lists 120![d] Quite a number of problems, some clearly of interest, remain obscure.

---

[d]I had copied Part 1 from the microfilm at the Warburg Institute. While there were some other interesting topological problems in Part 2, I only copied a few pages of these. I found it difficult to read the Italian (many words are run together and/or archaic) and the diagrams referred to are lacking. When I did work on Part 2, I found Warburg had mislaid the microfilm. Dario Uri was able to carry on and found the Chinese Rings and about a dozen other examples of earliest known topological puzzles.

Figure 6.5. Picture of the truncated icosahedron, from the printed version.

6.2 Recreational Material in *De Viribus Quantitatis*

The folio numbers from the MS, the image number in Uri's photos and the pages in Peirani's transcription are given. The numbered problems are not discussed in the order they appear.

Part 1

First European mention of "Blind Abbess and Her Nuns."

89. "Of an abbot who tries to guard a certain monastery of monks in the Levant by counting evening and morning the same on each side and how the sneering desperados abandoned it." *De uno abate ch' tolse aguardar certo monasterio de monache in levante contandole sera e matina per ogni verso tante et pur daloro schernito desperato la bandona* [ff. IVv–Vr = Uri 12 = Peirani 8.].

Figure 6.6. Portrait of Pacioli and the Brera Altarpiece by Piero, c. 1472.

This is the problem where there is a 3×3 square with three nuns in each exterior cell and the Blind Abbess can only count the number of nuns along each side, namely nine. But this allows considerable variation in the number of nuns present. At least one Arabic version is known.

The use of weights

85. [Part 3] "To make four weights which weigh to 40." *De far 4 pesi che pesi fin 40* [f. XIIIv = Uri 29 = Peirani 20].

The Indice for the third part has this but at the end Pacioli says this problem is in "libro nostro," i.e. the *Summa* [5] (p. 6). In the *Summa* [ff. 97r–97v, no. 34] Pacioli gives a general discussion of the use of 1, 3, 9, 27, 81, 243, ..., as weights.

86. [Part 3] "Of 5 cups of diverse weights to pay the landlord every day." *De 5 tazze, diversi pesi ogni di paga l'oste* [f. XIIIv = Uri 29 = Peirani 20].

Again this problem is in "libro nostro" [5] (p. 6). In the *Summa* [ff. 97v–98r, no. 35] Pacioli gives the problems of using five cups to pay daily rent for 30 days using the cups of weight 1, 2, 4, 8, 15.

Perfect numbers

26. "Effect to find a number thought of if it is perfect." *Effecto a trovare un nū pensato quando sia perfecto* [ff. 44v–47r = Uri 117–122 = Peirani 74–77].

Pacioli's discussion of perfect numbers has an amusing error. He gives the first five perfect numbers as 6, 28, 496, 8128, 38836. The last is actually $4 \times 7 \times 19 \times 73$ and is very wrong but it is clear in the MS. We do have $38836 = 76\ M_{11}$, so it seems Pacioli erroneously thought $M_{11} = 511$ was prime, but the multiplication by 256 was corrupted into multiplication by 76, probably by shifting the partial product by 2 into alignment with the partial product by 5.

First One Pile Game

34. "Effect to finish whatever number is before the company, not taking more than a limiting number." *Effecto afinire qualunch' numero na'ze al compagno anon prendere piu de un termi(n)ato.n* [ff. 73v–76v = Uri 175–181 = Peirani 109–112].

The One-Pile Game is a Nim variant, except with just one pile and a limit on the amount one can play. Early versions were usually additive. Here the players can add a number less than 7 to a pile and the object is to achieve 30. He describes how to win

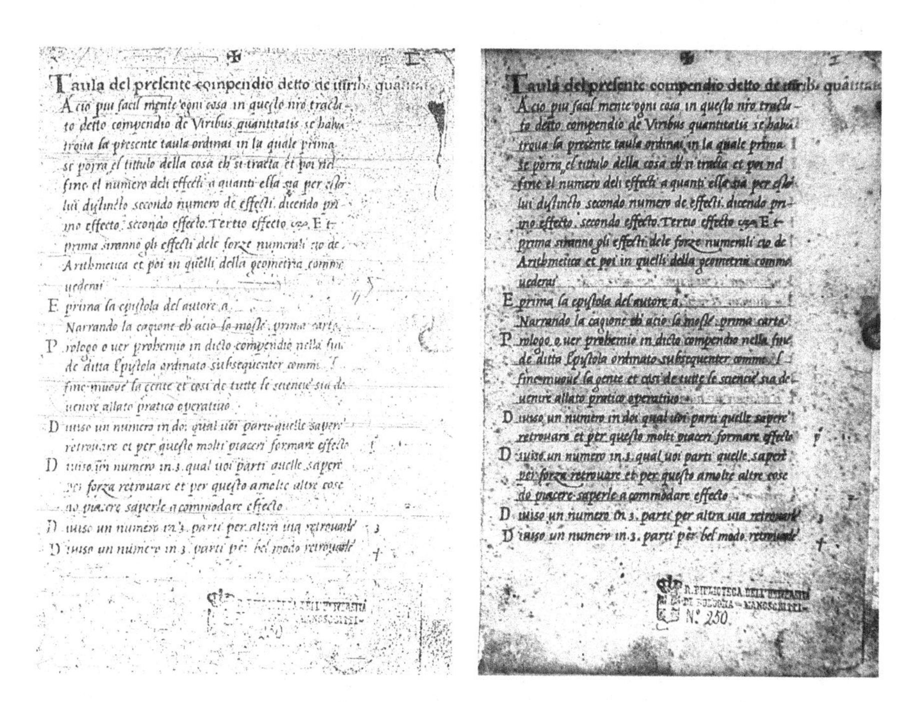
Taula del presente compendio detto de uirib. quantitatis
A cio piu facil mente ogni cosa in questo nro tractato detto compendio de Viribus quantitatis se habia
troua la presente taula ordinata in la quale prima
se porra el titulo della cosa ch' si tracta et poi nel
fine el numero deli effecti a quanti ella sia per esso
lui distincto secondo numero de effecti. dicendo primo effecto. secondo effecto. Tertio effecto &c. E t
prima se narranno gli effecti dele forze numerali cio de
Arithmetica et poi in quelli della geometria comme
uederai
E prima la epistola del autore a.
Narrando la cagione ch' acio la mosse. prima carta
Prologo o uer prohemio in dicto compendio nella fine
de ditta Epystola ordinato subsequenter comme l
fine muoue la gente et cosi de tutte le sciencie sia de
uenire allato pratico operatiuo
Diuiso un numero in doi qual uoi parti quelle sapere
retrouare et per questo molti piaceri formare effecto
Diuise un numero in .3. qual uoi parti quelle saperle
per forza retrouare et per questo amolte altre cose
de piacere saperle a commodare effecto
Diuiso un numero in 3. parti per altra uia retrouarle
Diuiso un numero in .3. parti per bel modo retrouarle

Figure 6.7. First page of the Indice, from the microfilm: F. Ir = Uri 4 = Peirani 3 and from Uri.

this and the general game. There are some early 16th century references which may be to Nim-type games, but this is the earliest, so we can view it as the ancestor of all Nim or take-away games!

Three odds make an even

He gives several problems of this type, which are impossible tricks. These only appear previously in Alcuin.

47. "A cashier who placed on the table some piles of ducats as a good trick." ff. 92v–93v = Uri 213–215 = Peirani 132–133. *de un casieri ch' pone in taula al quante poste de d(ucati) aun bel partito* [ff. 92v–93v = Uri 213–215 = Peirani 132–133].

Place four piles each of 1, 3, 5, 7, 9 ducats. Ask the person to take 30 ducats in 5 piles. If he can do it, he wins all 100 ducats. Discusses other versions, including putting 20 pigs in 5 pens with an odd number in each. However, the Italian word for 20, *vinti* (written *uinti*, can be divided into five parts as u–i–n–t–i, and each part is one letter.

48. "About another who placed some other even piles, good trick." *Ch' pur unaltro pone al quante altre poste pare bel partito* [ff. 93v–94r = Uri 215–216 = Peirani 133–134].

Place four piles each of 2, 4, 6, 8, 10 *carlini* (a small coin of the time) and ask the person to take 31 *carlini* in 6 piles.

The Indice lists the above as problems 50 & 51 and lists problem 52:

52. "On the dubious placing of 30 pigs in 7 odd pens." *Del dubio amazar 30 porci in 7 bote disparre* [f. IIIr = Uri 8 = Peirani 6].

133. [Part 3] "Tell me how to divide 'vinti' into five odd parts." *Dimme come farrai a partir vinti in 5 parti despare* [*F.* 281v = Uri 591 = Peirani 407, no. 133].

Divides as v–i–n–t–i, and mentions dividing 20 into 7 pens.

First optimal solutions for the Crossing the Desert Problem

The only earlier example of this problem is in Alcuin and Alcuin fails to find the optimum solution. Pacioli does four examples, finding the best solution each time.

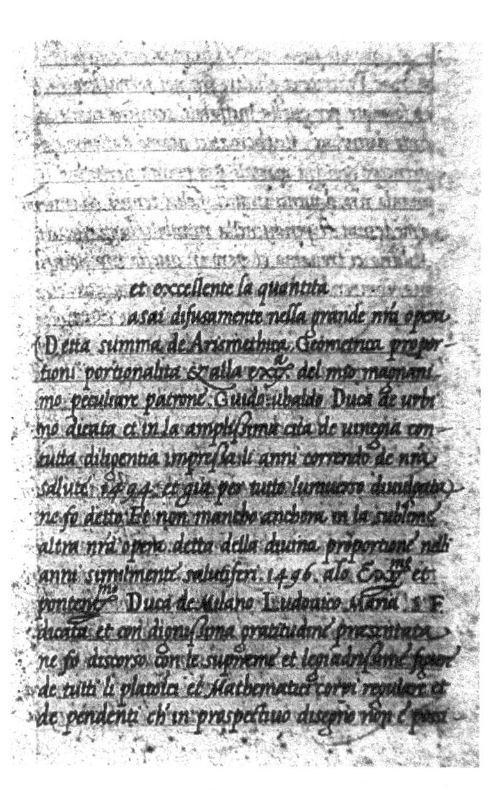

et excellente la quantita
asai difusamente nella grande nra opera
Detta summa de Arismethica Geometria proportioni portionalita & alla exc.a del mio magnanimo peculiare patrone Guido ubaldo Duca de urbino dicata et in la amplissima cita de venegia con
tutta diligentia impressa li anni correndo de nra
salute 1494. et gia per tutto luniverso divulgata
ne fo detto. Et non mancho anchora in la sublime
altra nra opera detta della divina proportione nelli
anni similmente salutiferi 1496. alo Ex.mo et
potent.mo Duca de Milano Ludovico Maria S F
dicata et con dignissima gratitudine presentata
ne fo discorso con le supreme et legiadrissime figure
de tutti li platonici et Mathematici corpi regulari et
de pendenti ch'in prospectivo disegno non e possi-

Figure 6.8. Title page: F. 15 = Uri 30 = Peirani 21.

49. "Of two ways to transport as many apples as possible." *de doi aportare pome ch' piu navanza* [ff. 94r–95v = Uri 216–219 = Peirani 134–135].

One has 90 apples to transport 30 miles from Borgo [= San Sepolcro] to Perosia [= Perugia], but one eats one apple per mile and one can carry at most 30 apples. He carries 30 apples 20 miles and leaves 10 there and returns, without eating on the return trip! (So this is the same as Alcuin's version.) Pacioli continues and gives the optimum solution!

50. "Of three ships holding 90 measures, passing 30 customs points." *de 3 navi per 30 gabelle 90 mesure* [f. 95v = Uri 219 = Peirani 136].

Each ship has to pay one measure at each customs point. Mathematically the same as the previous.

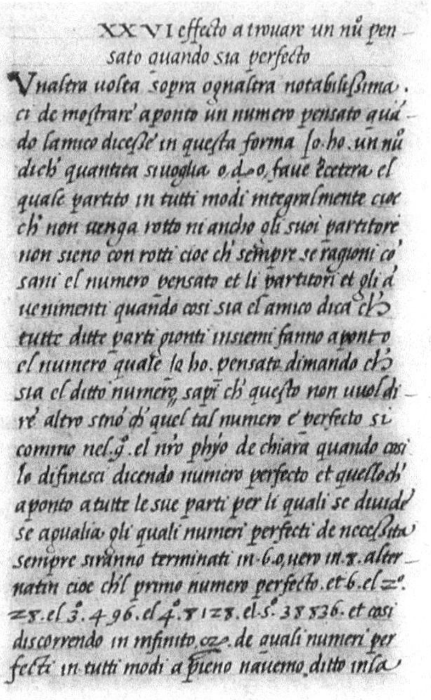
XXVI effecto a trovare un nũ pen-
sato quando sia perfecto
Vnaltra uolta sopra ognaltra notabilissima.
ci de mostrare aponto un numero pensato quã-
do lamico dicesse in questa forma io ho. un nũ
dicħ quantita siuoglia o d. o, faue &cetera el
quale partito in tutti modi integralmente cioe
cħ non uenga rotto ni ancho gli suoi partitori
non sieno con rotti cioe cħ sempre se ragioni cõ
sani el numero pensato et li partitori et gli a-
uenimenti quando cosi sia el amico dica cħ
tutte ditte parti gionti insiemi fanno aponto
el numero quale io ho. pensato dimando cħ
sia el ditto numero sapi cħ questo non uuol dire altro sinõ cħ quel tal numero e perfecto si
commo nel.9°.el nr̃o phyo de chiara quando cosi
lo difinesci dicendo numero perfecto et quello cħ
aponto a tutte le sue parti per li quali se diuide
se agualia. gli quali numeri perfecti de necessita
sempre sirãnno terminati in.6.o,uero in.8.alter-
natim cioe cħl primo numero perfecto.et 6.el 2°.
28.el 3°.496.el 4°.8128.el 5°.32536.et cosi
discorrendo in infinito, &c. de quali numeri perfecti in tutti modi a pieno hauemo ditto in la

Figure 6.9. F. 44v = Uri 117 = Peirani 74.

51. "To carry 100 pearls 10 miles, 10 at a time, leaving one every mile." *de portar 100 perle 10 miglia lontano 10 per volta et ogni miglio lascia* [f. 96r = Uri 220 = Peirani 136–137].

Takes them 2 miles in ten trips, giving 80 there. Then takes them to the destination in 8 trips, getting 16 to the destination.

52. "The same with more carried by another method." *el medesimo con piu avanzo per altro modo* [ff. 96v–97r = Uri 221–222 = Peirani 137].

Continues the previous problem and takes them 5 miles in ten trips, giving 50 there. Then takes them to the destination in 5 trips, getting 25 to the destination.

[This is optimal for a single stop — if one makes the stop at distance a, then one gets $a(10-a)$ to the destination. One can make more stops, but this is restricted by the fact that pearls cannot be divided. Assuming that the amount of pearls accumulated at each depot is a multiple of ten, one can get 28 to the destination by using depots at 2 and 7 or 5 and 7. One can get 27 to the destination with depots at 4 and 9 or 5 and 9. These are all the ways one can put

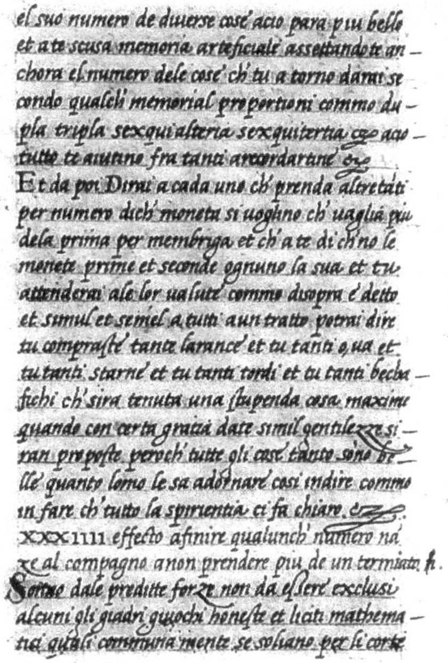

el suo numero de diuerse cose acio para piu belle
et a te scusa memoria arteficiale assettandote anchora el numero dele cose ch' tu a torno darai se
condo qualchi memorial propertioni commo du-
pla tripla sexquialteria sexquitertia &c acio
tutto te aiutino fra tanti arecordarane &c
Et da poi Dirai a cada uno ch' prenda altretati
per numero dech' moneta si uoglino ch' uaglia piu
dela prima per membriga et ch' a te di ch'no le
monete prime et seconde ognuno la sua et tu
attenderai ale lor ualute commo disopra é detto
et simul et semel a tutti aun tratto potrai dire
tu comprasti tante larance et tu tanti oua et
tu tanti starne et tu tanti tordi et tu tanti becha-
fichi ch' sira tenuta una stupenda cosa maxime
quando con certa gratia date simil gentilezze si-
ran proposte peroch' tutte gli cose tanto sono belle quanto lomo le sa adornare cosi indire commo
in fare ch' tutto la spirientia ci fa chiaro. &c
XXXIIII effecto afinire qualunch' numero na-
ze al compagno anon prendere piu de un terminato fi.
Sonno dale predItte forze non da essere exclusi
alcuni gli giadri giuochi honeste et liciti mathematici quali communa mente se soliano per le corte

Figure 6.10. F. 73v = Uri 175 = Peirani 108–109.

in two depots with integral multiples of 10 at each depot and none of these can be extended to three such depots. If the material being transported was a continuous material like grain, then perhaps the optimal method is to first move 1 mile to get 90 there, then move another $\frac{10}{9}$ to get 80 there, then another $\frac{10}{8}$ to get 70 there, ..., continuing until we get 40 at 8.4563 ..., and then make four trips to the destination. This gets 33.8254 to the destination. Is this the best method?]

First Impossible Jug Problem

55. "Of two other subtle divisions of bottles as described." *de doi altri sotili divisioni. de botti co'me se dira* [ff. 98v–99r = Uri 225–226 = Peirani 139–140].

Given a bottle of size A full of wine, divide it in half using two bottles of sizes B and C. After several genuine examples, he gives $A, B, C = 10, 6, 4$ and 12, 8, 4. Pacioli suggests giving these to idiots. This kind of impossible problem is actually rare — I've only noted two other examples.

First Josephus Problem counted out to the last two

The Josephus or "counting-out" problem appears in European MSS back to the 9th century and also appears in Japan, possibly as early as the 12th century and clearly from c. 1331. The classical version has 30 passengers on a ship, of two types which we will label "good guys" and "bad guys" — 15 of each. A fierce storm arises and the captain announces that half of the passengers must go overboard to save the ship. Someone suggests that they all stand in a circle and count out every ninth person, who then has to go overboard, willingly or not. After each departure, the count continues, going around the reduced circle. Surprisingly (or not), it happens that all the bad guys go overboard.

The early Japanese versions are the first known examples of counting to the last man. In 1539, Cardan introduced this idea into Europe and suggested this was how Josephus had escaped death. Josephus was a Jewish captain in the rebellion of the Jews against the Romans from 66 CE. He and forty of his fellow citizens were hidden under the city of Jotapata as it was overrun by Vespasian. He urged the men to surrender, but they preferred to die, and chose lots, each man striking off the head of the previously chosen man. The standard version of Josephus's text says he survived "by chance or God's providence" but a Slavonic version says he "counted the numbers with cunning and thereby misled them all." Josephus went on to become a historian of the Jews and the Jewish War, but he gives no further details.

56. "Of Jews and Christians in diverse methods and rules, to make as many as one wants, etc." *de giudei Chri'ani in diversi modi et regole. a farne quanti se vole etc* [ff. 99r–102r = Uri 226–232 = Peirani 140–143].

Pacioli gives six versions of the problem as problems 56–60 (with an unnumbered problem after 56) [ff. 99r–103v = Uri 226–235 = Peirani 140–146]. In three problems — 56, unnumbered, 57 — he leaves two survivors, which is the first time that this occurs. Unusually, there is a marginal diagram by the first problem showing the process. Two good guys & thirty bad guys counted by nines. (The marginal diagram is in Figure 6.11, but it is not in the transcription and Peirani says another diagram is lacking.)

Pacioli suggests counting the passengers on shore and doing the counting out with coins or pebbles in case one will need to know the

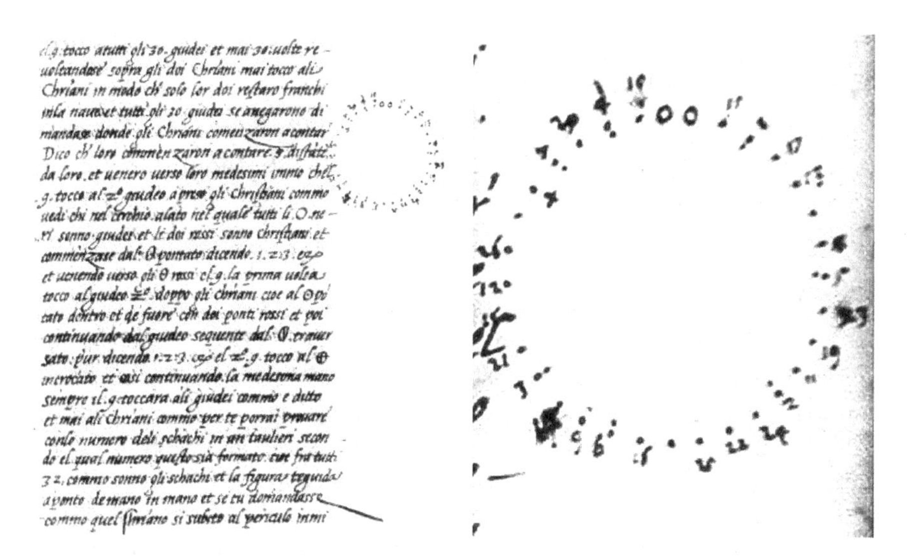

Figure 6.11. F. 73v = Uri 175 = Peirani 108-109 and an enlargement of the diagram on this page.

arrangement in a hurry. He also says one might count by 8s, 7s, 6s, 13s, etc., with any number of Christians and Jews. In examining this, the unexpected feature is that the two survivors, marked by circles at the top, were adjacent in the original circle. This seemed most unlikely, but one soon sees that the same behavior holds for counting out 31, 30, 29, ..., 3 by nines.[e]

The other problems which leave two survivors have counting 2 & 18 by 7s and counting 2 & 30 by 7s. In the first case, the survivors are adjacent in the original circle, but not in the second case. Neither has a diagram.

The unnumbered problem. *De 18 Giudei et 2 Chri'ani* [f. 102v = Uri 233 = Peirani 144].

57. "Of 30 Jews and 2 counting by 7 with the touched going in the water." *De 30 Giudei et 2 contando per 7 ch' toca va in aqua* [ff. 102r–102v = Uri 232–233 = Peirani 144].

---

[e]I found this sufficiently intriguing that I have written a paper on how to determine the largest n such that counting out by ks leaves two adjacent survivors; "Adjacent survivors in the Josephus Problem," submitted to *Mathematics Magazine*.

First River Crossing Problems with more jealous husbands or with larger boats

61. "About 3 husbands and 3 wives." *De .3. mariti et .3. mogli gelosi* [ff. 103v–105v = Uri 235–239 = Peirani 146–148].

Solution for 3 couples. Says that 4 or 5 couples requires a 3 person boat.

Octagram Puzzle

An early version of this puzzle. This has an octagram (or an octagon) and one has to place seven counters on it by placing each counter on a point and moving it ahead one place (or three places). The earlier versions were the problem of shifting seven knights located on the edges of a 3×3 board, which is known from c. 1275.

68. "Of a city with 8 gates which admits of rearrangement." *D(e). cita ch' a .8. porti ch' cosa convi(e)ne areparарli* [ff. 112r–113v = Uri 252–256 = Peirani 158–160].

Octagram puzzle with a complex story about a city with 8 gates and 7 disputing factions to be placed at the gates. The Indice gives the above as problem 83. And 82 seems likely to describe a similar problem.

82. "Of 8 ladies who are at a ball and of 7 youths who accompany them." *De 8 donne ch' sonno aun ballo et de 7 giovini quali con loro sa con pagnano* [f. IVv = Uri 11 = Peirani 8].

First Western Binary Divination

69. "To find a coin thought of among 16." *A trovare una moneta fra 16 pensata* [ff. 114r–116r = Uri 256–260 = Peirani 161–162].

Divides 16 coins in half 4 times, corresponding to the value of the binary digits. Pacioli does not describe the second stage clearly, but Agostini makes it clear. This idea is supposed to have been common in Japan from the 14th century or earlier, but examples have not yet been seen. Pacioli gives many other simple divinations, some based on the Chinese Remainder Theorem and the classic problem of divining a permutation of three items.

First Rearrangement on a Cross

This is a variation of the Blind Abbess and her Nuns, where a person has a cross and counts the jewels on it from the base to each other end. A clever jeweler or pawnbroker removes some jewels.

la 3.a cioe.c. peroch' questi .2. ultimi fili sono
sira .g. et .c. a .q.o.m.k. laltro sia .h.f.d.b.
.r. p.n.l. et cosi obseruarai sempre et date
piu altre ne proportionarai &c.
CAp. LXX. D. un prete ch' in pegno la
borscia del corporale con la croci de p(er)le
al Giudeo
QVuanto facia al huomo acolto scaltrito la for-
za et uirtu deli numeri oltra legia asegnate
occurrentie el fa manifesto el caso ch' gia a-
un certo pouano interuene el quale a certi
suoi bisogni no hauendo altri denari ando
al giudeo a impegnare una bella ueste de
corporale de gran ualuta per una certa q.a
de d. sopra la quale eranno alquante gro-
sse perle de stima situate in croci la qual croci
era in questo modo facta commo uedi qui in
margine et cioe ch' contandole al piedi cioe da-
piedi al capo eranno noue perle et gli piedi
con luu de bracci anchora .9. et cosi gli piedi
con laltro braccio similmente .9. et lasciandola
al giudeo acio non si scambiasse ne prese el
ditto contrase deman del giudeo cioe ch' da lui
receue una ueste tale &c. con una croci tale

Figure 6.12. F. 116r = Uri 260 = Peirani 162–163.

70. "Of a priest who pledges to a Jew the burse of the corporale with a cross of pearls." *D(e). un prete ch' in pegno la borscia del corporale con la croci de p(er)le al Giudeo* [ff. 116r–117v = Uri 260–263 = Peirani 162–164].

Fifteen jewels with three on each arm, one counts to nine from the base to each arm end. This is reduced to thirteen. Asks how one can add one pearl and produce a count of ten — answer is to put it at the base.

Magic Square

Pacioli associates magic squares with planets and gives Dürer's magic square of 1514, but both of these had been done before.

72. "Of numbers arranged in a square by astronomers, which total the same in all ways, along sides and along diagonals, as symbols of the planets and suitable for many puzzles and how to put them." *D(e). Numeri in quadrato disposti secondo astronomi ch' p(er) ogni verso fa'no tanto cioe per lati et per Diametro figure de pianeti et*

amolti giuochi acomodabili et pero gli metto [ff. 118r–118v, 121r–122v (some folios are wrongly inserted in the middle) = Uri 264–265, 270–273 = Peirani 165–167].

Gives magic squares of orders 3 through 9 associated with planets in the system usually attributed to Agrippa (1533), but this dates back to at least the early 14th century. There are spaces for diagrams [ff. 121v and 122r], but they are lacking. He gives the first two lines of the order 4 square as 16, 2, 3, 13; 5, 10, 11, 8; so it is the same square as given by Dürer.

Selling different amounts "at the same prices" yielding the same

65. "Of a merchant who has three agents and sends them to a market with pearls." *De un mercante ch' a 3 factori et atutti ma'da auno mercato con p(er)le* [ff. 119r–119v = Uri 266–267 = Peirani 154–155].

In addition, there are four more examples listed in the Indice [ff. IIIv–IVr = Uri 9–10 = Peirani 7] as problems 70–73.

70: "Of another merchant who sends three agents to a fair with varying numbers of pearls and they sell them at the same price and they each carry as many pence as the others to the master at home." *De unaltro mercante ch' pur a 3 factori et mandali a una fiera con varia quantita de perle' et vendano a medesimo pregio et portano acasa tanti denari al patrone uno quanto laltro.*

71: "Of another variant of the preceding with three agents having various quantities of pearls at equal prices and likewise take as many pence to the master." *De unaltro vario dali precedenti ch' pur a 3 factori con vari quantita de perle' pregi pari et medesimamente portano al patrone d(enari) pari.*

72: "Of another merchant who has 4 agents to whom he gives various numbers of pearls which they sell at the same prices and receive equal money." *De unaltro mercante ch' ha 4 factori ali quali da quantita varie di perli ch' amedisimi pregi le vendino et denari equalmente portino.*

73: "Of another who sends 4 agents with varying numbers of pearls and they report back to the house the same prices and the same money, variation of the preceding." *De un altro ch' pur a 4 factori con quanti(ta) varie di perle apari pregi et pari danari reportano a casa vario dali precedenti.* I have studied this problem in: Some diophantine recreations; IN: *The Mathemagician and Pied Puzzler A Collection in Tribute to Martin Gardner*; ed. by Elwyn R. Berlekamp & Tom Rodgers; A. K. Peters, Natick, Massachusetts, 1999, pp. 219–235.

Combining amounts and prices incoherently

This type of problem is sometimes called the "Applesellers' Problem" or the "Marketwomen's Problem."

66. "Of one who buys 60 pearls and resells for exactly what they cost and gains." *D. de uno ch' compra 60 perle et revendele aponto per quelli ch' gli stanno et guadago* [ff. 119v–120r = Uri 267–268 = Peirani 155–156].

Buy 60 at 5 for 2, sell 30 & 30 at 2 for 1 and 3 for 1.

The Indice [f. IVr = Uri 10 = Peirani 7] lists the above as problem 74 and continues with

75: "Of another merchant who buys 60 pearls at a certain price for a certain quantity of ducats and resells them at the same price at which he bought them and gains a ducat but with different effort than the preceding." *De unaltro mercante ch' pur compro perle' .60. a certo pregio per certa quantita de ducati et sile ceve'de pur al medesimo pregio ch' lui le comparo et guadagno un ducato ma con altra industria dal precedente.*

Gathering apples from a garden

67. "A master who sends a servant to gather apples or roses in a garden." *Un signore ch' manda un servo a coglier pome o ver rose in un giardino* [ff. 120r–120v, 111r–111v (some pages are misbound here) = Uri 268–269 & 250–251 = Peirani 156–158].

(Lose half and one more) three times to leave 1. Discusses the problem in general and also does (Lose half and one more) five times to leave 1; (Lose half and one more) three times to leave 3.

Collecting Stones

He gives an example which is a simple summation of an arithmetic progression.

73. "To pick up 100 stones in a line." *D(e). levare 100 saxa a filo* [ff. 122v–124r = Uri 273–276 = Peirani 167–169].

Wager on the number of steps to pick up 100 stones (or apples or nuts), one pace apart. Gives the number for 50 and 1000 stones.

Constructing (approximate) n-gons

Pacioli gives rules for $n = 9, 11, 13, 17$. Let L_n denote the side of a regular n-gon inscribed in a unit circle. Nick Mackinnon [6] has studied these.

23. "To make the 7th figure called nonagon, that is of 9 sides, difficult." *Afare la 7a fia dicta nonangolo. cioe de 9 lati difficile* [ff. 147r–147v = Uri 322–323 = Peirani 198–199].

Asserts $L_9 = \frac{L_3+L_6}{4}$. Mackinnon computes this gives 0.6830 instead of the correct 0.6840.

25. "On the 9th rectilinear figure called undecagon." *Documento della 9 fia recti detta undecagono* [ff. 148r–148v = Uri 324–325 = Peirani 200].

Asserts $L_{11} = \phi\frac{L_3+L_6}{3}$ where $\phi = \frac{1-\sqrt{5}}{2}$, is the "golden mean." Mackinnon computes this gives 0.5628 instead of the correct 0.5635.

26. "On the 13th." *Do. de' .13* [f. 148v = Uri 325 = Peirani 200].

Asserts $L_{13} = \frac{5}{4}(1-\phi)$. Mackinnon computes this gives 0.4775 instead of the correct 0.4786.

28. "On the 17-angle, that is the figure of 17 sides." *Documento del 17 angolo cioe fia de 17 lati* [ff. 149r–149v = Uri 326–327 = Peirani 201–202].

Peirani says some words are missing in the second sentence of the problem and Agostini says the text is too corrupt to be reconstructed.

Part 2

First Staircase Cut

79. "To know how a tetragon can be lengthened with contraction, enlarged with shortening." *Do(cumento). un tetragono saper lo longare con restregnerlo elargarlo con scortarlo* [ff. 189v–191r = Uri 407–410 = Peirani 250–252].

Since this problem has an added diagram and seemed to use a trick cut, perhaps it is an ancestor of the vanishing area puzzles. Pacioli's description is a little cryptic and is thoroughly confused by an erroneous diagram added at the bottom of the page [f. 190v, redrawn on Peirani 458] — this must have been added by a reader who did not understand the phrasing. Once one realizes what is going on, the text is reasonably clear. He is converting a 4×24 rectangle to a 3×32 using one cut into two pieces. So this is the common

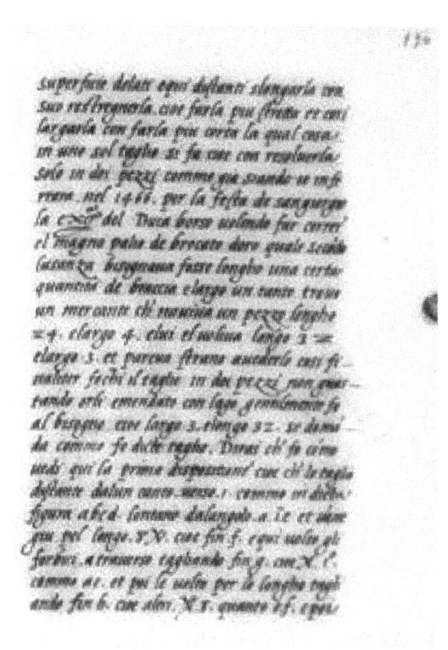

Figure 6.13. Left: F. 190v = Uri 409 = Peirani 251–252, without the diagram. Right: F. 190r = Uri 408 = Peirani 251.

problem $4A \times 3B$ to $3A \times 4B$ with $A = 1$, $B = 8$, which is done by a "staircase" cut giving two pieces which can be assembled into a second rectangle. Below the diagram [f. 190v] is an inserted note which Peirani [252] simply mentions as difficult to read, but some bits of it are legible. The drawing and the note suggests he made a cut and then moved one piece so the cut would continue through it to make three pieces with one trick cut. Pacioli clearly notes that the area is conserved.

First Place Four Points Equidistantly

80. "How it is not possible for more than three points or discs or spheres to all touch in a plane." *Do(cumento) commo non e possibile piu ch' tre ponti o ver tondi spere tocarse in un piano tutti* [ff. 191r–192r = Uri 410–412 = Peirani 252–253].

This is a bit vaguely described. It says you can only get three discs touching in the plane, but you can get a fourth so they are all touching by making a pyramid.

Pacioli gives the earliest known versions of six "topological puzzles." Unfortunately, only one of these has a picture, though they

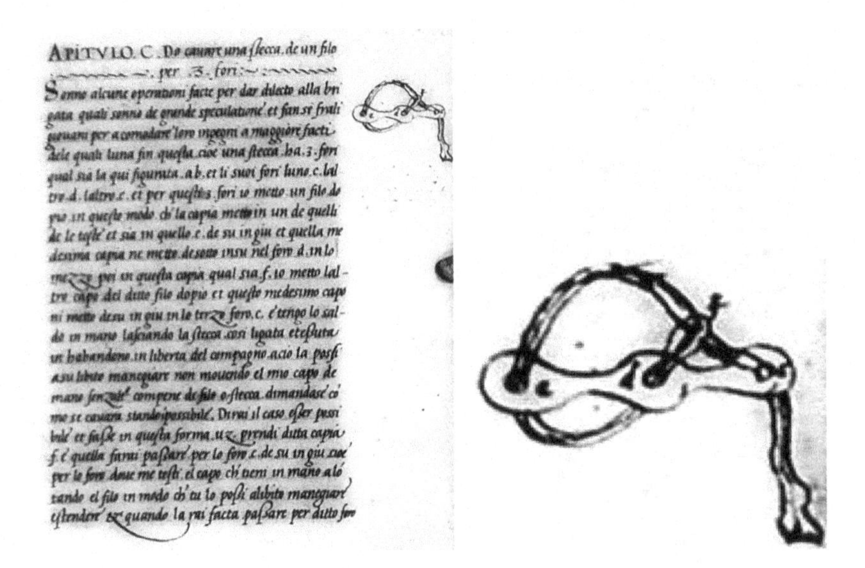

APITVLO. C. Do cavare una stecca. de un filo per .3. fori:

Senne alcune operationi facte per dar dilecto alla brigata quali sonno de grande speculatione et fanse fra li [illegible] per acomodare loro ingegni a maggiore facti. dele quali luna sin questa cioe una stecca .ha.3.fori qual sia la qui figurata .a b. et li suoi fori luno.c.lal tro.d. laltro.e. et per questi .3. fori io metto un filo do pio in questo modo. ch' la capia metto in un de quelli de le teste et sia in quello .e. de su in giu et quella me desima capia ne metto de sotto insu nel foro d. in lo mezzo [illegible] in questa capia qual sia .f. io metto lal tre capi del ditto filo dopio et questo medesimo capo ni metto desu in giu in le terzo foro.c. e tengo lo sal do in mano lasciando la stecca cosi ligata et [illegible] [illegible] in liberta del compagno acio la possa a sui libito maneggiare non movendo el mio capo de mano senza [illegible] compere de filo o stecca. dimandase co me se cavara stando impossibile. Dirai il caso esser possi bile et fasse in questa forma .u[illegible]. prendi ditta capia f. e quella farai passare per lo foro .c. de su in giu cioe per le foro dove me testi el capo ch' tieni in mano alo tando el filo in modo ch' tu lo possi al libito maneggiare estendere et quando la rai facta passare per ditto foro

Figure 6.14. F. 206r = Uri 440 = Peirani 282-283, without the diagram. Detail from this page.

generally refer to one! Dario Uri has greatly extended the number of these.

First Victoria Puzzle

In the late 19th century, this was called the "Victoria" or "Alliance Puzzle."

100. "To remove a stick from a cord through 3 holes." *Do(cumento) cavare una stecca. de un filo per .3. fori* [ff. 206r–206v = Uri 440–441 = Peirani 282–283].

There is a marginal drawing [f. 206r] clearly showing the string through three holes in one stick, but this is not reproduced in the transcription. Uri has found that several further problems are describing similar puzzles.

102. "Another speculation — remove two buttons from a string divided in the middle and halved at the ends." *Do(cumento) unaltro speculativo cavar doi botoni di una stenga fessa nel mezzo et sce'pia in testa* [ff. 207v–208v = Uri 443–445 = Peirani 284–286].

Dario Uri says this is describing a version of the Victoria or Alliance puzzle with four holes in each button.

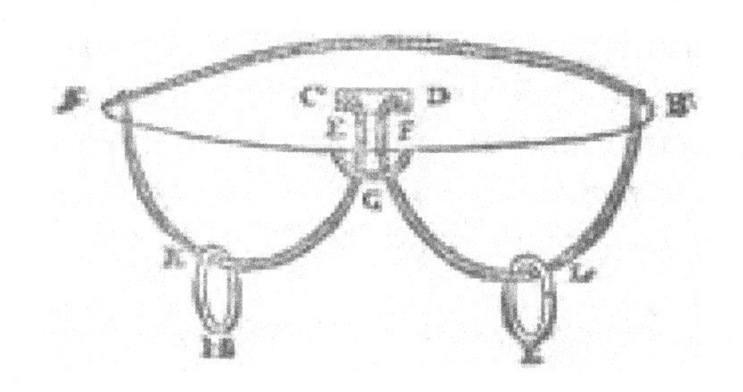

Figure 6.15. Solomon's Seal from Schwenter, 1636.

103. "To tie two shoe soles together into one with the above mentioned doubled string — a beautiful thing." *Do(cumento) legare con la sopra detta strenga fessa. doi sola. de carpe' ambe doi. a uno modo. bella cosa* [ff. 209r–210r = Uri 446–448 = Peirani 286–288].

Pacioli says this is *quasi simile alla precedente.* Dario Uri illustrates this with the Alliance or Victoria puzzle from Alberti (later as Figure 8.4).

108. "Remove a large ring from two tied to a stick by the ends." *Do(cumento). Cavare' uno anello grande fore' de doi legati a una bacchetta per testa* [ff. 213r–215r = Uri 454–458 = Peirani 292–295].

Dario Uri says this is a version of this idea and illustrates it with an unidentified picture.

First Solomon's Seal

101. "Another string also through three holes in the stick with one bead per loop, make them go onto one (loop)." *Do(cumento) un altro filo pur in 3 fori in la stecca con unambra. per sacca far le andare' tutte in una* [ff. 206v–207r = Uri 441–442 = Peirani 283–284].

The problem titles vary between the actual problem and the Table of Contents and the latter shows that *unambra* should be *una ambra* — Peirani has given it as *un'ambra. Sacca* means pocket or bay or inlet and it seems clear he means a loop which has that sort of shape. *Ambra* is amber, but seems to mean an amber bead here.

First Cherries Puzzle

This has two versions and Pacioli gives both.

104. "To remove and replace two cherries in a cut card." *Do(cumento) cavare' et mettere' .2. cirege' in una carta tramezzatta* [ff. 210r–210v = Uri 448–449 = Peirani 288–289].

Pacioli's description clearly shows there is one hole, but Dario Uri illustrates this with the picture from Alberti which has two holes.

110. "A button from a (cross)bow or two cherries from a button and bow." *Do(cumento) uno bottone' de un balestro. o vero doi cirege' de un botone' et balestro* [ff. 215v–216v = Uri 459–461 = Peirani 296–297].

Dario Uri translates *balestro* as "flexible stick" and illustrates it with Figure 8.5.

First European Interlocking Chinese Rings

This may be the first description of this in the world.

107. "Remove and replace a joined string a number of joined rings — a difficult thing." *Do(cumento) cavare et mettere una strenghetta salda in al quanti anelli saldi. dificil caso* [ff. 211v–212v = Uri 451–453 = Peirani 290–292].

Dario Uri found this describes the Chinese Rings. It has seven rings. On his website, Uri gives several of the legends about its invention and says Cardan called it Meleda, but that word is not in Cardan's text. He lists 27 patents on the idea in five countries. It is supposed to have originated in China, but definite evidence is lacking.

First Three Knives Make a Support

129. "Join together three blades of knives." *Do(cumento) atozzare .iij tagli de coltelli insiemi* [ff. 228r–228v = Uri 484–485 = Peirani 315].

Pacioli says this was shown to him on 1 April 1509 (Peirani has misread 1509 as isog) by *due dorotea veneti et u perulo 1509 ad primo aprile ebreo.* Peirani transcribes *u* as *un* but Dario Uri thinks it is the initial of Perulo's given name. Perhaps *dorotea* refers to some occupation, e.g. nuns at S. Dorothy's Convent? In Vienna, the Dorotheum is a huge public auction house where estates are auctioned off. The word *ebreo* means "Hebrew," but it is unclear what it refers to.

First Jacob's Ladder

The early form with two boards is also known as the Chinese Wallet or Flick-Flack. Versions with more boards appear in the late 18th century.

九连环和巧环——由阮根全制作
Ingenious Rings Puzzles by Ruan Genquan

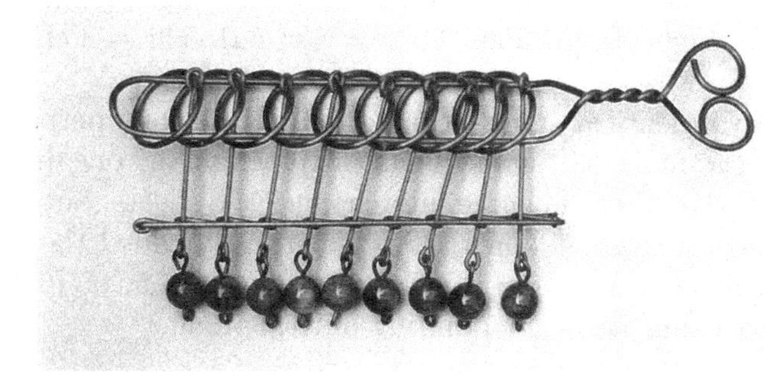

九连环—要将框柄从所有九个连环中套出需 341 步
Nine Interlocking Rings is a classical Chinese puzzle.
Removing the handle from all nine rings requires 341 moves.

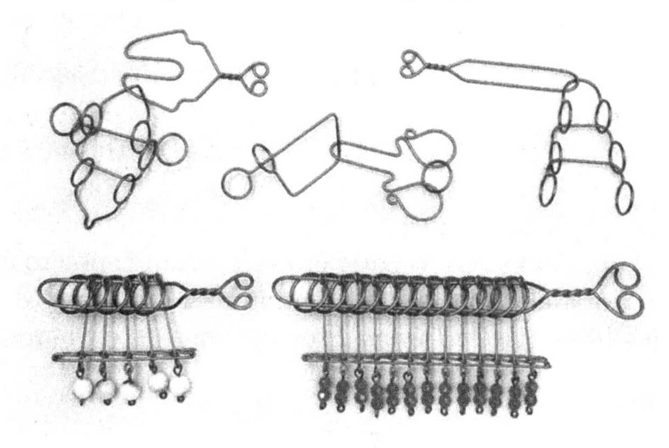

双套连环、剪和刀环、十全环、五连环、十三连环
Five Interlocking Rings,Thirteen Interlocking Rings,and other puzzles

Figure 6.16. Some modern examples of Interlocking Rings from China.

132. "On the childish recreation called deception." *Do(cumento) del solazo puerile ditto bugie* [ff. 229r–229v = Uri 486–487 = Peirani 316–317].

Uses two tablets and three leather straps. Describes how to use it to catch a straw. His references show that it is called *calamita di legno* [calamity of wood] (or magnet of wood — depending on whether the Italian is *calamità* or *calamita*).

Part 3

First discussion of a Geometrical Optical Illusion

73. "To deceive someone's eyes, an illusion." *Do. in gannare' uno della vista abagliarlo* [ff. 256v–257r = Uri 541–542 = Peirani 364–365].

Takes two identical strips of paper and places one perpendicular to the other to make a T shape. 9 out of 10 people say one direction is longer than the other. Then he interchanges the sheets, but the same direction is still seen as longer. This is generally called the vertical–horizontal illusion and is attributed to Oppel, 1855, or Fick, 1851. (Thanks to Vanni Bossi for pointing out this item.)

First Two Fathers and Two Sons Make Only Three People

He gives several other "strange family" puzzles [f. 287v, no. 191 = Uri 603 = Peirani 416].

Vanni Bossi and Bill Kalush have looked at the tricks and discovered several earliest examples of magic tricks — see Bossi's contribution which follows this. Those who have looked at this now feel this is definitely the earliest recreational mathematics book — except that it was never published.

A statue of Pacioli was recently erected in San Sepolcro in the park at Via Matteotti and Via de gli Aggiunta, see Figure 6.18. The base uses designs from his *De Divina Proportione* — see Figure 6.18. The

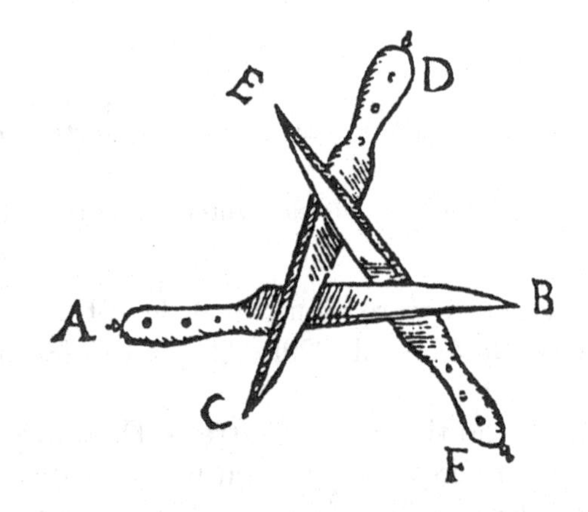

Figure 6.17. Three knives from Prevost, 1584.

Figure 6.18. Statue of Luca Pacioli in Borgo Sansepolcro.

leading hotel/restaurant in town is the Albergo Ristorante Fiorentino at Via Luca Pacioli 60 and the proprietor is interested in Pacioli — in the restaurant is a banner from the 500th anniversary celebrations of the publication of *Summa* in 1494. There is a plaque with his portrait on the Palazzo delle Laudi.

Bibliography

[1] PACIOLI-SUMMA.
[2] PACIOLI-DIVINA.
[3] PACIOLI-DVQ.
[4] PIERO.
[5] A. Agostini. "De viribus quantitatis di Luca Pacioli." *Periodico di Matematiche* 4, 4 (1924) 165–192.
[6] N. Mackinnon. "The portrait of Fra Luca Pacioli." *Mathematical Gazette* 77, 479 (July 1993) plates 1–4 & pp. 129–219).

Chapter 7

Pacioli's Magic and Card Tricks

by Vanni Bossi†

The *De Viribus Quantitatis* manuscript is a collection of "ludi mathematici" (mathematical games or recreations) where the author wishes to teach mathematics avoiding weariness of repeated exercises that normally ask for the power of intellect and patience.[a]

For the same reason, other authors before him did the same (like Leonardo Pisano, better known as "Fibonacci," or both Francesco and Pier Maria Calandri). But, in other *trattati d'abbaco* (such as [2])[b] recreational problems are placed here and there in the text, just to give a "pause" to the mind, while Paciolo's manuscript, instead, can be considered a real treatise on the subject.

The relationship with magic is clearly understandable in some of Paciolo's statements; for instance, he gives great importance to the secret of the method used to accomplish the effect to astonish the viewers (this is a basic and fundamental principle in magic). So great is his care in hiding the secret, because the secret is the

[a] A version appeared in: Demaine, Erik D.; Demaine, Martin L. & Rodgers, Tom; *A Lifetime of Puzzles Honoring Martin Gardner*; A. K. Peters, 2008, pp. 77–122. This is the text of Bossi's contribution to our joint presentation on *De Viribus Quantitatis* at the Gathering for Gardner 6 (2004). I have added some explanatory notes. My grateful thanks to all the friends, especially to Bill Kalush, Mark Setteducati and P. G. Varola.

[b] Abacus treatises, which were actually arithmetic textbooks; both *abbaco* and *abaco* occur.

conditio sine qua non to be able to amaze your friends, especially — he says — *i rozzi* [the rude fellows] and *maxime donne* [especially women], who do not know mathematical principles because neither had access to the *scuole d'abaco* [abacus schools or schools of arithmetic]. So, if some rules are easy to learn and master, some others are intentionally complicated, augmenting the number of necessary operations to avoid detection. This allows you both to give a different presentation of the same effect, or to disguise the method using a different one; and this is a second fundamental principle greatly used in magic. Further, in most cases the spectator is invited to simply think of a number or the value of a coin or card, which adds to the mystery.

Luca Pacioli's (or Paciolo, or Fra Luca da Borgo Sansepolcro or "fra Luca") manuscript has remained unpublished for about 500 years. We could consider the Peirani-Marinoni transcription as the first printed edition of this work [2]. While unpublished, we have much evidence that it has certainly been a source for later works.[c] Fra Luca does not claim originality; some of the games come from older works. Some have been invented by his students and proper credits are given (Paciolo explains that he encouraged his students to do this).[d]

According to Gilberto Govi, one of the great scholars of Leonardo of the 19th century, most of the tricks should be credited to Leonardo. This is absolutely possible, as Fra Luca taught him arithmetic and Euclid's geometry, although unfortunately he does not give credit to him for any of them. It is known that two *giochi di partito* [party games] are described in notebooks of Leonardo, so this could be a proof of the connection. Again Govi wrote that Leonardo was performing a trick where some light objects made of wax (my note:

[c]One popular work is Bachet [1] to which, in the past, various scholars attributed the priority for being the first work on recreational arithmetic. Bachet's respected work is the first book printed but the honor of being the first collection of entertaining mathematical problems belongs to Pacioli.

[d]For instance, he names in Chapter XLVII his disciple Carlo de Sansone from Perugia: in Chapter XLVIII he names Catano de Aniballe Catani from Borgo who performed the game in Naples in 1486. This date is interesting because it points out that most of the problems in the manuscript could have been invented in the last quarter of the 15th century.

probably tissue waxed to make it stiff and moldable) were made to fly in the way oriental performers make paper butterflies fly![e]

Back to Paciolo, we can also suppose that he shared some of his secrets with street conjurers or court performers; this can explain his knowledge of most of the non-arithmetical tricks and puzzles described in the second and third parts of the manuscript. This could also explain the finding of some principles explained in the manuscript (that we know it was never printed and hardly accessible to common people) in many pamphlets of "secrets" usually sold by itinerant performers. These secrets were probably transmitted orally and occasionally printed.[f]

Coins, dice and cards are the objects most used in the explanation of tricks based upon arithmetical principles, the reason being that all these objects can represent a number or quantity (an amount of equal coins with each one a unity, or coins with different values; dice with six faces of different values and thc possibility to use more than one of them; playing cards, with values from 1 to 13, and the possibility of creating combinations thanks to color and suit). Coins and dice had been used before (i.e. in Calandri's work) but the use of cards is described for the first time in Paciolo's work.

None of the effects are necessarily executed with cards, but in some instances he says that the same trick done with cards is more deceptive. Another interesting thing is the justification that Paciolo, being a Franciscan Monk, gives to the use of such objects as dice,

[e]Soon after the death of Professor Augusto Marinoni, Giunti Editore in Florence published a monumental work that is his complete transcription of Leonardo's "Codice atlantico." I met Prof. Marinoni many times (he was living in Legnano very close to where I live) and he confirmed to me that Leonardo had an interest in conjuring and was performing some tricks. Unfortunately, the personal notes and files of Prof. Marinoni are unpublished and unavailable at present.

[f]Many of these booklets have been discovered recently. As far as is presently known, the most extensive work is Horatio Galasso's *Giochi di carte bellissimi, di regola e di memoria ...* published in Venice, 1593, where the author describes 25 card tricks (including the first printed system for a stacked deck), many of them based on arithmetical principles, some of which can be found in Paciolo. This is followed by 25 "secrets" of various types, some of which can also be found in Paciolo. Of this booklet I have made a reprint with an introduction in 2001. It has been translated into English and I hope it will soon be available to English speaking people.

cards and *trionphi* (tarot cards), usually considered of an unbecoming nature for a religious person: he uses them not to gamble but to demonstrate the power of numbers in an easily comprehensible manner.

Let us see now in which chapters he describes card magic:

"Fourth effect: of a number divided in three parts, etc.," *Quarto effecto: de un numero in tre parti diviso, etcetera* [1] [Peirani-Marinoni 30].

He gives the description of this effect with numbers, dice and cards.

"Fifth effect: of a number divided by 4, or in 4 parts," *Quinto effecto: de un numero diviso fra 4, o vero in 4 parti* [Peirani-Marinoni 36].

This is an interesting principle, still in use today. He assigns mentally a number to each of four spectators, each one holding a different card. With the rule given, you know who has which card.

"30th effect: of a number thought of, multiplied several times, the products by the same or different numbers, to find the resulting division," *XXX effecto: de numero pensato, multiplicato pi volte gli suoi producti per diversi o medesimi numeri, trovare l'avenimento partito* [Peirani-Marinoni 87]. Less literally: to find a number thought of, from the result of its being multiplied several times, by the same or different numbers.

In this chapter, Paciolo describes the use of a confederate, a child, that secretly holds a paper where all the results of the possible multiplications are written; and when the performer asks for a product, the child is instructed how to answer properly. To make this easier for the child, and also more impressive, he suggests putting the child in another room. In this way, the child can easily read directly from the paper without being seen. Then Paciolo suggests that the sequence of the tricks should follow a path with high and low, thus obtaining more emotional involvement; practically the rules of a theatrical performance. He also says that by instructing the child, it is possible to secretly communicate to him by code words, or gestures, or signals (cough, tapping with a knife on the table and so on) or by numbers. He mentions a magician whose name was Jasone da Ferrara who was performing such effects with a boy in gentleman's houses where Paciolo was personally attending.

"35th effect: to know how to find three different things distributed among three person, and four distributed among four, and of as many as you wish," *XXXV effecto: de saper trovare 3 varie cose divise fra tre persone, et 4 divise fra 4, et de qua(n)te vorrai* [Peirani-Marinoni 112].

This chapter is mentioned, although no card tricks are described, because Paciolo suggests to memorize the operations by verses, a mnemonic method that will be later widely used (and still is) to remember the order of the sequence of the cards in a stacked deck.

"Chapter 40: to know of two things, distributed one in each hand or between two people or two unequal numbers, one even and one odd, without any questioning," *XL Capitolo: de doi cose una per mano divise o ver fra doi, o ver doi numeri inequali, paro et imparo, senza alcuna interrogazione sapere* [Peirani-Marinoni 118].

In this chapter a method for guessing between two spectators is given, who has an odd and who has an even number of things. In the second example, two cards, one odd, one even, are used. They are thrown on the table face up, so you can see their value.

"Chapter 64: [to know] of a number thought of by means of a circle," *d'un numero pensato per via de un cerchio* [Peirani-Marinoni 151].

In this trick, cards are the perfect things to use. It is the classical "clock trick" where a known number of cards are placed in a circle on the table, face down (you only know the value and the order of them, that is progressive). Now a spectator is invited to think of a number not higher than twelve (if you are using twelve cards); then by given instructions he has to count starting from a point and he will finish on a card that when turned over will have the value of the thought number. This trick is described also in the Galasso booklet.

"Chapter 80: Of persons who instantly determine [a number] by natural means without other calculation," *D(e) le gentileza che a le volte si fanno per vie naturali senz'altro calcolo* [Peirani-Marinoni 177].

This chapter is mentioned, where no card tricks are found, since a very interesting principle is explained. Paciolo names two persons both using this principle: one is Francesco de la Penna and the other the already mentioned Giovanni de Jasone da Ferrara. The technique is what we today call in card magic "estimation." The performer knows by experience how many nails, or walnuts or whatever can fill a certain container. In performance he invites a spectator to fill for instance a bottle with nails and predicts how many will fill it with a very close approximation that amazes the audience. In the same way, he can say how many walnuts can be held in a fist and so on.

No more card tricks are described in the manuscript but a lot of magic and amusing physical principles as well as puzzles can be found.

The second part of the manuscript is largely devoted to geometry, but from Chapter 94 [Peirani-Marinoni 275] on, he describes a series of puzzles, some hydraulic principles as well as some optical recreations, some stunts based on physics, secret writings and a few magic tricks.[g]

The third part has no card tricks but is filled with very interesting things. Just to name the most intriguing:

Chapter 8 (second paragraph) [Peirani-Marinoni 334]: using prepared pieces of paper (some of which float on water, some not) you can make a friend become the victim that will pay a penalty.

Chapter 9 [Peirani-Marinoni 334]: Paciolo describes the right to left mirror writing of Leonardo which can be read with a mirror.

Chapter 10 [Peirani-Marinoni 335]: how to write a sentence on the petals of a rose or other flowers.

Chapter 11 [Peirani-Marinoni 335]: how to engrave letters on iron by the use of chemicals (this technique will be developed soon afterward for engraving and printing).

Chapter 23 [Peirani-Marinoni 342]: a method for washing the hands in molten lead without being hurt (a stunt used by mountebanks and fireproof performers).

Chapter 36 [Peirani-Marinoni 348]: how to cut a pigeon's neck with a knife without killing him. And in the next chapter:

Chapter 37 [Peirani-Marinoni 349]: how to kill a pigeon by hitting his head with a feather ... this time really killing him!

Chapter 40 [Peirani-Marinoni 350]: how to make an egg crawl on a table. Into a hollowed egg is placed a leech. The hole in the egg is closed with wax. Placing the egg next to a vase filled with water, the leech will feel the water and start to move to reach it, making the egg crawl on the table. A similar method (using a bug) will be

---

[g] Many of these are described later in Cardano's and Della Porta's works.

found described in many pamphlets of secrets of the 16th and 17th centuries.

Chapter 41 [Peirani-Marinoni 350]: a very "modern" method to make a coin go up and down into a glass filled with water using powder of "calamita" (magnetite).

Capitolo XLIIII [Peirani-Marinoni 352]: another effect of a coin dancing into a glass on your command using a woman's hair attached at one end to the coin with wax, the other end, again with wax, attached to your finger (a method still in use).

Chapter 46 [Peirani-Marinoni 353]: how to eat tow [yarn] and spit fire. A very old classic trick, still in use today, by street performers and pseudo-fakirs.

Chapter 66 [Peirani-Marinoni 362]: how to cut a glass spiral shaped so that it works as a spring.

Chapter 73 [Peirani-Marinoni 364]: an optical illusion. [This is demonstrating the inability to compare horizontal and vertical distances.]

Bibliography

[1] BACHET.
[2] PACIOLI-DVQ.
[3] PIERO.

Chapter 8

Some Early Topological Puzzles

Often the 1723/1725 edition of Ozanam [6] is considered to be the first book to cover topological puzzles in detail, with only a few earlier examples. Ozanam certainly gives many more examples than any previous book. However, we discuss some sources for several topological puzzles that are considerably older. This chapter assumes some familiarity with these puzzle types.

The most important development in the field is that Dario Uri has been systematically examining Part Two of Pacioli [7], see Chapter 6 which is difficult to understand. He has discovered that the following puzzles occur: the Alliance or Victoria Puzzle, Solomon's Seal, the Cherries Puzzle, the Chinese Wallet, and the Chinese Rings! Further, Chapter 109 is the problem of joining three castles to three wells by paths that do not cross; Chapter 117 is a trick of removing a ring from a loop between a person's thumbs [9] (p. 410, the top of Figure 12); Chapter 121 is the trick where strings are just looped inside a ball so they can pulled away; Chapter 129 is the problem of making a support from three knives. As far as we know, these are the earliest appearance of all these puzzles. And there are about 140 problems in this part and Uri has not done them all. Unfortunately, though Pacioli refers to diagrams, the only diagram in the MS is for Chapter 100. Perhaps Uri will find the missing drawings in the library at Bologna. Uri has found that Chapter 129 refers to Pacioli being shown the problem on 1 April 1509, so perhaps this MS should be dated as 1510.

8.1 The Chinese Wallet or Flick-Flack or Jacob's Ladder

In May 1998, Peter Hajek reported seeing a painting at Hampton Court showing this — see Figure 8.1. Some research in art history books turned up a picture of it and a reference to another depiction, probably earlier.

This is a painting by the fairly well known Lombard follower of Leonardo, Bernardino Luini (c. 1470–1532). The picture is variously cited as "A Boy with a Toy" or "Cherub with a Game of Patience." There is no indication of its date, but the middle of Luini's working life is c. 1520. A description says the tapes holding the boards together are red and are apparently holding a straw, but does not seem to recognize the object. The painting was exhibited in London in 1898 and a contemporary review in a German journal calls it a

Figure 8.1. The Luini Painting and the Luini Engraving.

"Taschenspielerstückchen," a little juggler's trick — but recall that juggler was long a synonym for magician — with two boards which allow one to vanish the straw.[a]

This picture is in the Picture Gallery at Hampton Court Palace, see Figure 8.2. It is attributed to Bernardino Licinio, a Venetian painter born about 1485 and last known in 1549. It is 22″ × 18″

Figure 8.2. Licinio Painting.

[a]Tancred Borenius & J. V. Hodgson. "A Catalogue of the Pictures at Elton Hall in Huntingdonshire in the possession of Colonel Douglas James Proby." The Medici Society, London, 1924, has an image of this, Plate 5. The facing p. 11 describes the picture and says the picture is 17" by $13\frac{1}{2}$" (43 by 35 cm). The same child appears in another Luini painting, Christ and the Baptist as Children. The authors identify the "straw" as a fishbone and say it will disappear. My thanks to Richard Mankiewicz, Double Plus Books, Peterborough, for letting me see and photograph the book. The engraved version is by F. Bartolozzi, c. 1793, and I photographed James Dalgety's example.

(56 cm × 45 cm). It is described and illustrated in John Shearman's *The Early Italian Pictures in the Collection of Her Majesty The Queen*; CUP, 1983, where it is called "Portrait of a Man with a Puzzle." It is very similar to another painting known to be by Licinio and dated 1524, so this is probably c. 1524 and hence a bit later than the Luini. The description says the binding tapes are red, as in the Luini, and both show something like a straw being trapped in the wallet, which suggests some connection between the two pictures, though it may just be that this toy was then being produced in or imported to North Italy and was customarily made with red tapes. On the toy is an inscription: *Carpendo Carperis Ipse* (roughly [Snapping snaps the snapper]), but Shearman says it definitely appears to be an addition, though its paint is not noticeably newer than the rest of the painting. Shearman says the toy comprises "three or more rectangles ...," though both paintings clearly show just two pieces.

A little later in the 16th century, we have the first known published version of the Chinese Wallet, [8]], see Figure 8.3 left. This is the first book devoted to magic and conjuring. It is rather rare — only five copies are known — and a 1987 facsimile is out of print. Fortunately, it has been recently translated into English. My thanks to Bill Kalush for bringing this to my attention. The figure is taken from the new edition [8] (1988 edition, pp. 136–140 with the diagram on p. 139).

Another recently received source is Schwenter's book of 1636, which has a version [9] (part 15, exercise 27, pp. 551–552), where he calls it "Ginmaul" — "yawning mouth" see Figure 8.3 right.

And another early source is Witgeest. However, most of the interesting material is not in the first edition of 1679 of which there is a recent facsimile. The new material apparently occurs in the 2nd ed. of 1682 [14] and this is so extensively revised and retitled as to constitute a new book. Jerry Slocum has supplied some photocopies from the 3rd ed. of 1686. There were many further editions, both in Dutch and in German. Much of the new material seems to be derived from Schwenter, but here it is clearly different; Witgeest [14] (1686, problem 66, 49–50) see Figure 8.4 left.

It is not clear when the toy advanced to having more boards and becoming the Jacob's Ladder. Ozanam does not give any version of the toy. An early example is a c. 1850 example of a "Hand operated game of changing pictures" illustrated in Daiken [3] (plate 6, p. 24).

Very Cunning Tablets Adorned with Strings of Different Colors, Under or Outside of Which You May, if You Wish, Make a Wand be Seen

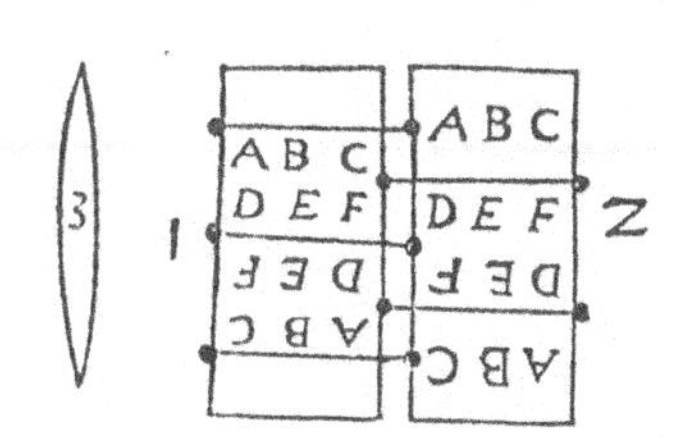

Die XXVII Auffgab.
Ein Ginmaul zu machen.
Die alten Künstler haben zwey Brättlein mit dreyen
künstlich zusamm geheffcet/daß sie/wann man das eine an die
gelt/das ander auff zweyen seiten daran hangen können/wie
einer Thür gesagt/welche an beeden Orten auff vnd zugehet/
sie ein Ginmaul genennet. Mach zwey viereckichte Brättlein/
fördern Spann lang/vnd vngefähr ein drittel der Läng bräit/
serrück dick a b c d, a e f c, in allem
darnach schneide drey schmale
Riemlein / welche auff beeden sei
Brättlein Farb haben / dann also
besser verdeckt/sie müssen aber alle
länger seyn/als ein Brättlein/das
einem ende neben an m auff das brä
mit dem andern an das l deß Brät

Figure 8.3. Prevost [8, p. 139] and Schwenter [9, p. 551].

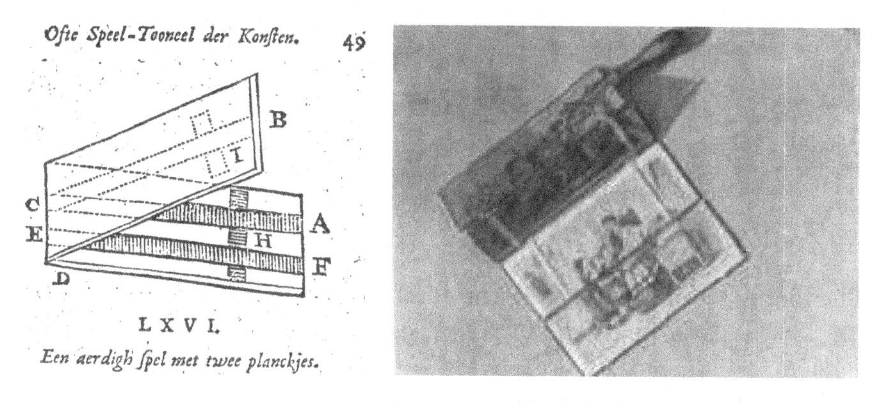

Figure 8.4. Witgeest [14, p. 49] and Daiken [3, p. 24].

One version has a Victorian lady on one side and an officer or nobleman on the other, undoubtedly a double portrait of Victoria and Albert. In the 20th century, this idea evolves into several puzzles — the tetraflexagon and Rubik's Magic.

8.2 The Alliance and Victoria Puzzle

This does occur in Ozanam [6] (vol. IV, problem 31, p. 435 & figure 37, plate 11 (13)), see Figure 8.5.

In 1557, Cardan shows both two hole and three hole versions [1] (vol. III, pp. 245–246), but Cardan's Latin is generally cryptic.

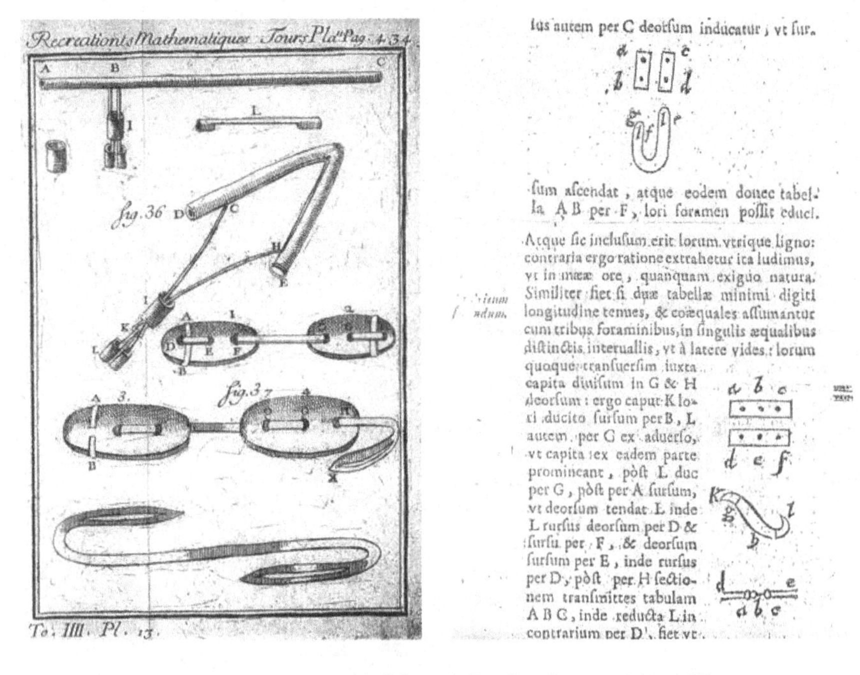

lus autem per C deorſum inducatur, vt ſur-

ſum aſcendat, atque eodem donec tabel-
la A B per F, lori foramen poſſit educi.

Atque ſic incluſum erit lorum vtrique ligno: contraria ergo ratione extrahetur ita ludimus, vt in mææ ore, quanquam exiguo natura. Similiter fiet ſi duæ tabellæ minimi digiti longitudine tenues, & coæquales aſſumantur cum tribus foraminibus, in ſingulis æqualibus diſtinctis interuallis, vt à latere vides: lorum quoque tranſuerſim iuxta capita diuiſum in G & H deorſum: ergo caput K lori ducito ſurſum per B, L autem per G ex aduerſo, vt capita ex eadem parte promineant, pòſt L duc per G, pòſt per A ſurſum, vt deorſum tendat L inde L rurſus deorſum per D & ſurſu per F, & deorſum ſurſum per E, inde rurſus per D, pòſt per H ſectionem tranſmittes tabulam A B C, inde reducta L in contrarium per D, fiet vt

Figure 8.5. Ozanam [6] and Cardan [1, pp. 245–246].

The pictures were repeated in the 1660 English version of Wecker [12] (p. 338. This is in Book XVIII.). See also Chapter 6 for Pacioli's version.

Both two-hole and three-hole versions occur in Prévost in 1584 [8, pp. 133–136], see Figure 8.6.

Schwenter gives a two hole version [9] (Part 10, exercise 29, pp. 410–411). Witgeest [14] (problem 44, pp. 35–36), is a two-hole version taken from Schwenter. Both are seen in Figure 8.7

8.3 Solomon's Seal or African Beads Puzzle

Let's start with the version in the 1723/25 Ozanam [6] (vol. IV, problem 40, pp. 439–440 & figure 47, plate 14 (16)). See Figure 8.8. It is called *Le Sigillum Salomonis, ou Sceau de Salomon.*

This is in Schwenter [9] (Part 10, exercise 27, pp. 408–410). Witgeest [14] (problem 43, pp. 33–34), Figure 8.9, is clearly taken from Schwenter. Pacioli's version is in Chapter 6.

To Enclose Two Tiny Pieces of Wood with Two Straps,[90] *So That One May Not Take Them Out Without Breaking the Wood or the Straps*

Have constructed two small, straight, and long pieces of wood, as you see drawn hereafter, each one having two round holes near the ends, which are marked with the four letters A, B, C, and D. Then take a slim leather strap of a

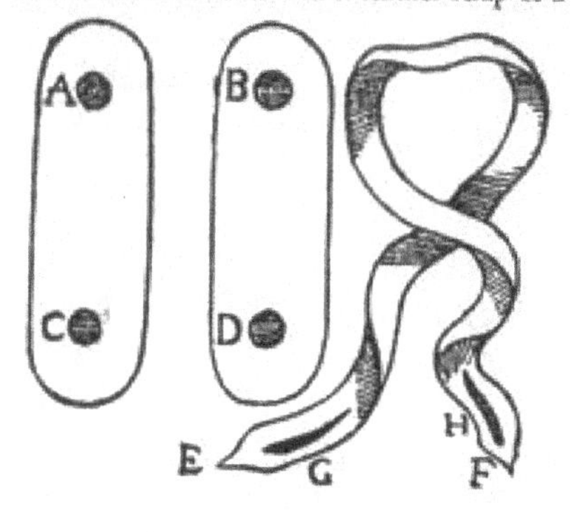

[133]

G through hole A, and end H through hole B, so that the two ends shall be on one side. Then pass end G through slit K and through hole C.

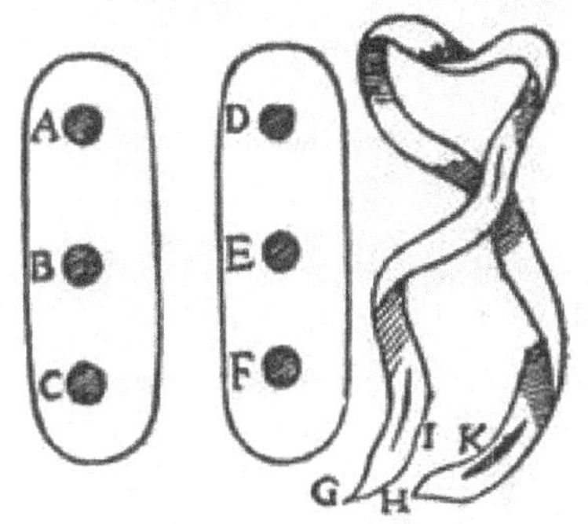

After this, to enclose the other piece, in a similar way pass end G through D and F, then through E and again through D, where the end is, holding the other piece, which you shall pass through slit I, and you shall pull through the strap from hole D. And the end shall be thoroughly caught in this piece as is the other end

[135]

Figure 8.6. Prevost [8, p. 133] and Prevost [8, p. 135].

Recently this has been called an African puzzle. A search revealed these references.

- R. P. Lelong. "Casse-tête guerzé." *Notes Africaines* 22 (April 1944) 1. Cited and described by Béart. Says M. Gienger found the variant with an extra ring encircling both loops in the forest of the Ivory Coast in 1940, named "kpala kpala powa" [body of a toucan] or "kpa kpa powa" [body of a parrot].
- Paul Niewenglowski. *Bulletin de l'IFAN* [Institut Français d'Afrique Noire] 14:1 (January 1952). Cited and described by Béart. Describes his invention of an interesting, rather simpler, variant as a result of seeing a standard version from Béart.
- Charles Béart. *Jeux et jouets de l'ouest africain.* Tome I. Mémoires de l'Institut Français d'Afrique Noire, No. 42. IFAN, Dakar, Senegal, 1955, Pages 413–418 discusses and carefully illustrates several versions. The standard version, but with several beads on

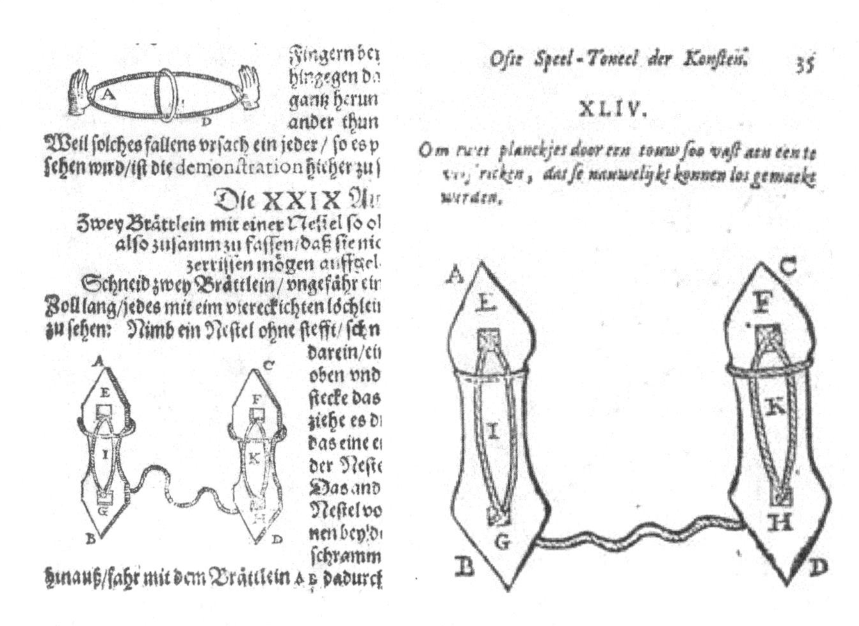
Fingern bey
hingegen da
gantz herun
ander thun
Weil ſolches fallens vrſach ein jeder / ſo es p
ſehen wird/iſt die demonſtration hieher zu ſ

Die XXIX Au
Zwey Brättlein mit einer Neſtel ſo ol
alſo zuſamm zu faſſen/daß ſie nic
zerriſſen mögen auffgel.
Schneid zwey Brättlein / vngefähr ein
Zoll lang/jedes mit eim viereckichten löchlein
zu ſehen: Nimb ein Neſtel ohne ſtefft/ ſteck n
darein/ein
oben vnd
ſtecke das
ziehe es d
das eine e
der Neſte
Das and
Neſtel vo
nen beyd
ſchramm
hinauß/fahr mit dem Brättlein A B dadurch

Ofte Speel-Toneel der Konsten. 35

XLIV.

Om twee planckjes door een touw soo vast aen een te vey'ricken, dat se nauwelijks konnen los gemaeckt worden.

Figure 8.7. Schwenter [9, p. 410] and Witgeest [14, p. 35].

one loop, is called pèn and is common in the forests of Guinea and Ivory Coast. It describes variants of Gienger/Lelong and Niewenglowski.

- Fred Grunfield. *Games of the World* (1975). On p. 267, he calls this "African String Game," but gives no reference.
- Pieter van Delft & Jack Botermans. *Creative Puzzles of the World.* Abrams, New York, 1978. African ball puzzles. "It was once used in magic rites by tribes living in the jungles of the Ivory Coast. The puzzle is still used for amusement in this part of Africa, not only by the people who inhabit the remote outlying areas but also by city dwellers. ... The puzzles were not restricted to this part of Africa. Variations may be found in Guinea, and some ... were made in China." No references are given.

8.4 The Cherries Puzzle

This comes in two common forms. Both are given in Ozanam. The classic cherries form is [6] (vol. IV, problem 33, p. 436 & figure 39, plate 12 (14)), see Figure 8.10.

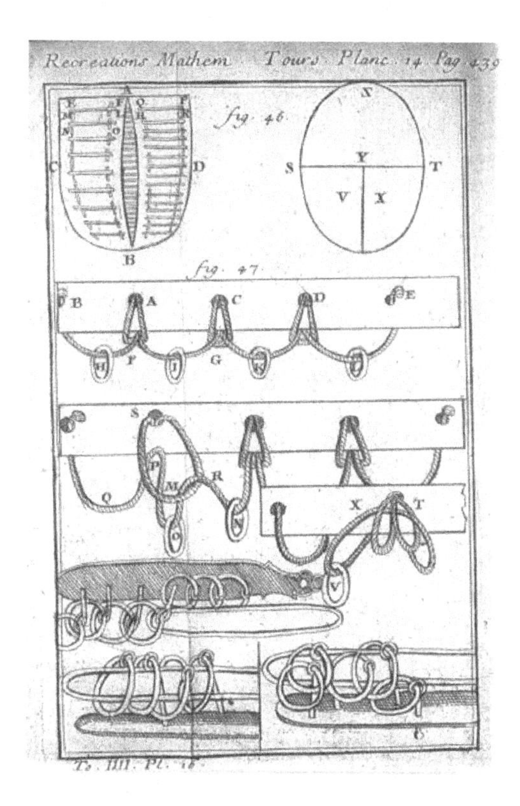

Figure 8.8. Ozanam [7] Problem 40.

Zehender Theil der Erquickſtunden. 409

gung mögen referirt werden/mir keins wunderſamer vorkommen/als diß/ obs zwar bey den Wiſſenden ein ſchlecht anſehen hat/wolte wündſchen/daß ich die demonſtration alſo dazu ſetzen köndte/ daß ſie von männiglich möchte verſtanden werden/weil ſie aber allzulang vnd mühſam/will ich den günſtigen Leſer damit nit moleſtiren oder beſchweren/ſondern einig vnd allein/ wie man hierinn practicire, jhme an die Hand geben. Ich halte dafür/daß niemand von ſich ſelbſt/beede Ring dem begeren nach/zuſamm bringen werde: Das Holtz aber dazu wird alſo gemacht: Nimb ein Holtz vngefehr eines Meſſerrucks dick/vnd einer ſpannen lang / ſpitze es zu wie bey der Figur im

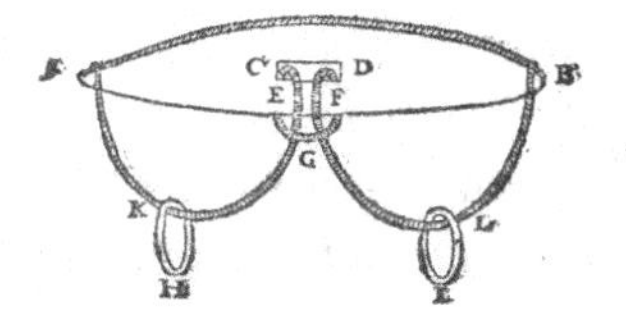

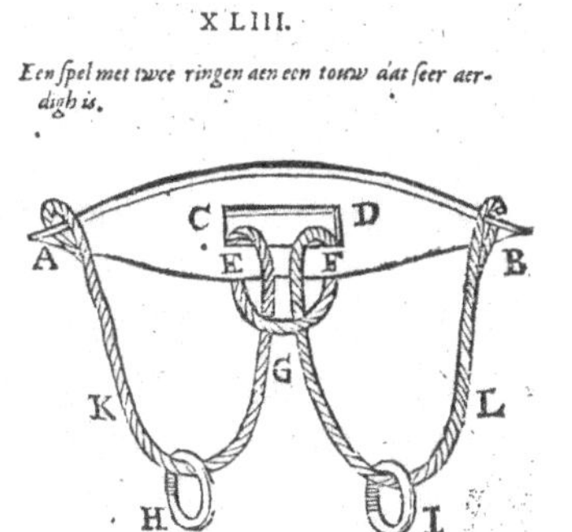

Ofte Speel-Toneel der Konſten. 33

X LIII.

Een ſpel met twee ringen aen een touw dat ſeer aerdigh is.

Figure 8.9. Schwenter [9, p. 409] and Witgeest [14, p. 33].

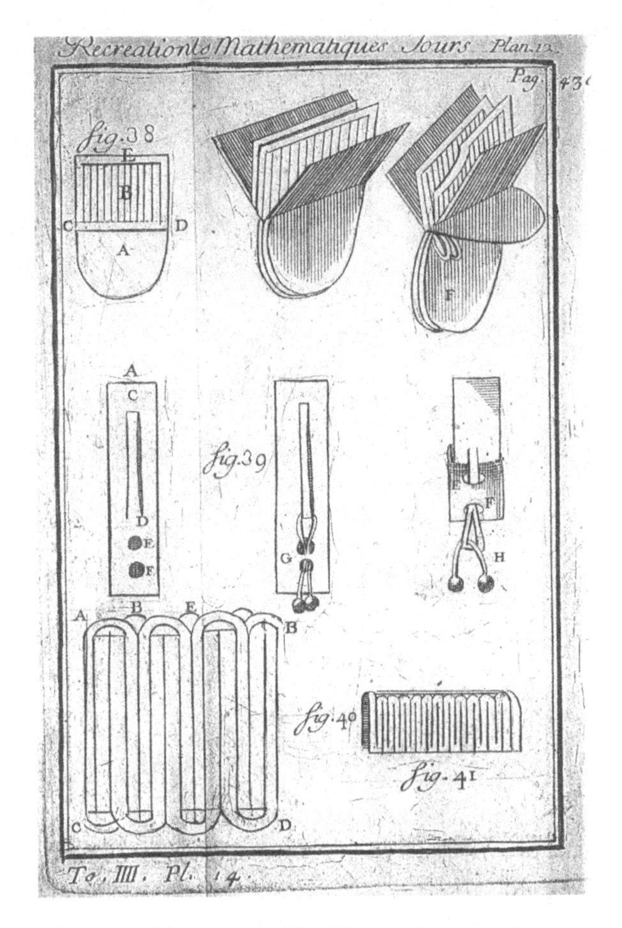

Figure 8.10. Ozanam [6] Problem 33. "On peut passer des queues de Cerises dans un papier, ...

Nothing earlier than Witgeest [14] (problem 19, pp. 162–163) has been found, who has a delightful picture with realistic cherries. Another pleasant picture comes from Minguét, the first magic book in Spanish, from 1733 [5] (pp. 112–113). Both are Figure 8.11.

The second version has a folded piece of paper or leather hanging from a card or tube which has two parallel cuts along much of it. The paper has large ends and there is a ring on the thin part which cannot go over the ends. The solution is essentially the same as for the Cherries Puzzle. We have already seen this as problem 36 in Figure 8.5.

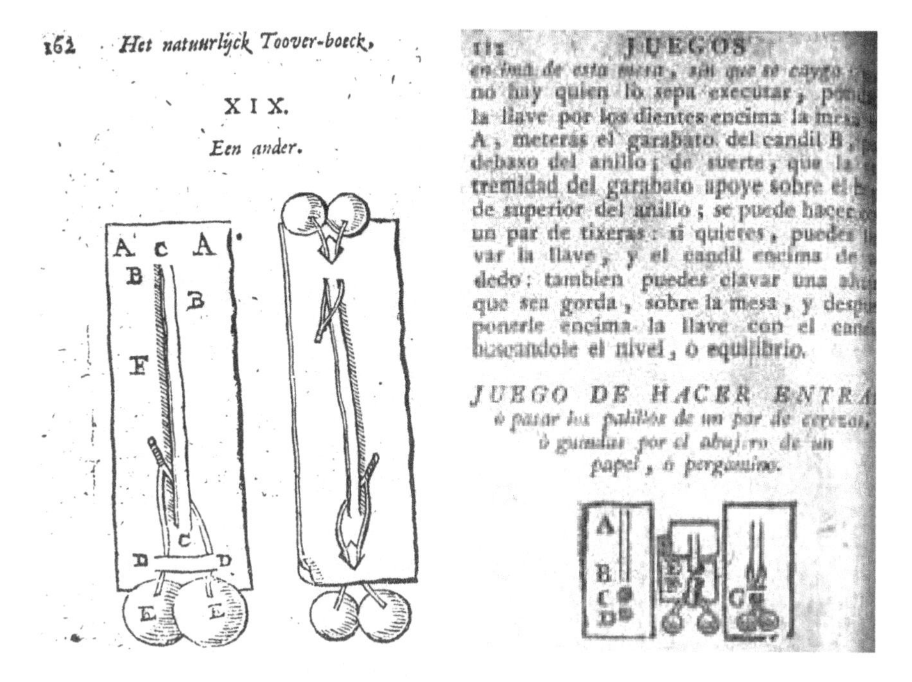

Figure 8.11. Witgeest [14, p. 162] and Minguét [5, p. 112].

We find a version in Schwenter [9] (part 10, exercise 30, p. 411). And a quite different version in Witgeest [14] (problem 18, pp. 160–161). Both are in Figure 8.12.

8.5 Six-Piece Burrs

Strictly this is not really topological. Unexpectedly a much earlier version was found. For some time, the example called "Die kleine Teufelsklaue" ("the little Devil's claw"), item 147 in the 1801 catalogue of Bestelmeier was the earliest known. But it turned up the 1790 catalog of the predecessor firm of Catel which has the same figures — see item 16 in Figure 8.14.

Dieter Gebhardt and Jerry Slocum then did a lot of hunting through German libraries and found the original parts of the Bestelmeier catalogue and early price lists of Catel, so we know the dates can be revised backward to 1794 and 1785, respectively [10]. However, the earliest known diagram of the pieces was in

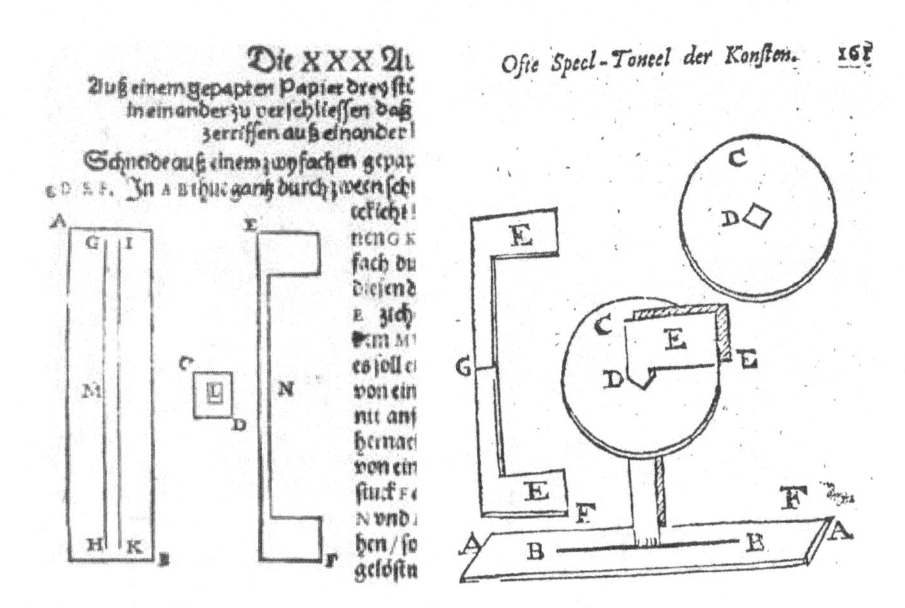

Figure 8.12. Schwenter [9, p. 411] and Witgeest [14, p. 161].

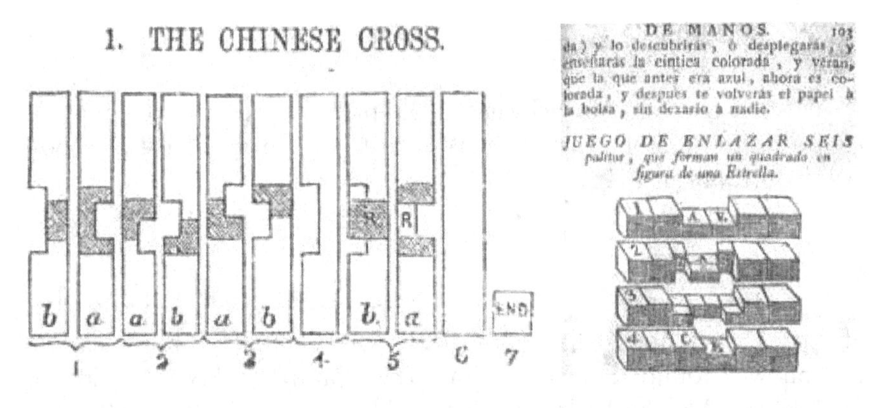

Figure 8.13. [4, p. 266] and Minguet [5, p. 103].

The Magician's Own Book of 1857, where it is called "The Chinese Cross" [4] (problem 1, pp. 266–267 & 291).

A clearer diagram, Figure 8.13, is in Minguét of 1733 [5] (pp. 103–105), where it is just called a star. Piece 3 must be duplicated and there is a plain key piece. This is not the same burr as the previous one.

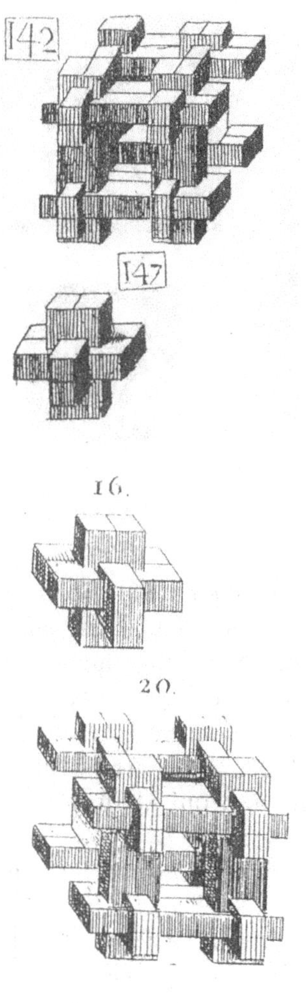

Figure 8.14. Bestelmeier-Catel.

8.6 Borromean Rings

This is the pattern of three rings where no two are actually linked, but all three are — see the bottom of Figure 8.15. It is part of the coat of arms of the Borromeo family, who are counts north of Milan since the 15th century. The Golfo Borromeo and the Borromean Islands are in Lago Maggiore. In the 16th and 17th centuries, the Counts of Borromeo built a baroque palace and gardens on the main island,

Figure 8.15. The Borromeo Crest and Japanese crest designs.

Isola Bella (or Isola Borromeo). The Borromean rings can be seen in many places in the palace and gardens, including the sides of the flower pots! Although the Rings have been described as a symbol of the Trinity, it is not clear how they came to be part of the Borromean crest. Perhaps the most famous member of the family was San Carlo Borromeo (1538–1584), Archbishop of Milan and a leader of the Counter-Reformation, but he does not seem to have used the rings in his crest.

Pietro Canetta. *Albero Genealogico Storico Biografico della nobile Famiglia Borromeo*, 1903, says it is copied from a manuscript of the archivist Pietro Canetta, with a footnote: Il Bandello, p. 243, vol. VIII. Probably this refers to a publication of the MS. This simply says that the three rings represent the three houses of Sforza, Visconti and Borromeo which are joined by marriages.[b]

Clarence Hornung, ed. *Traditional Japanese Crest Designs* (Dover, 1986). On plates 10, 20, 24 & 39 are examples of Borromean rings, shown in Figure 8.15.

These designs have no descriptions and the only dating is in the Publisher's Note which says such designs were common from the 12th to the 17th centuries.

[b]Thanks to Dario Uri for this source.

8.7 Chinese Rings

This is well known and undoubtedly known to the reader. There are several Oriental stories about its origins, but without any definite evidence to support these stories. One of the earlier versions is in Ozanam [6] (vol. IV; unnumbered figure on plate 14 (16)). Interestingly, there is no text corresponding to this picture.[c]

Until Dario Uri's recent discovery, the earliest known version of the puzzle was given by Cardan in 1550 [2]. Note Cardan's diagram of a single ring! In the 1663 *Opera Omnia*, the ring is stretched and labelled "Navicula" (little boat).

The first-known person to analyse the problem was John Wallis [11] and he gives several illustrations, notably Figure 8.16. Wei Zhang produced a new edition of a Chinese work at the time of the

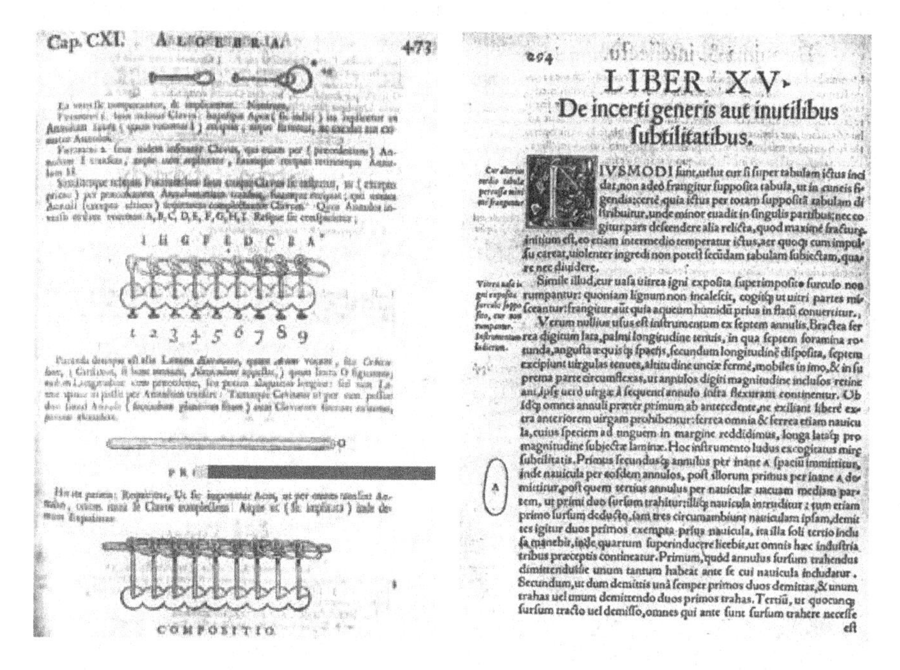

Cap. CXI. ALGEBRA. 473

I H G F E D C B A

1 2 3 4 5 6 7 8 9

COMPOSITIO

294

LIBER XV.

De incerti generis aut inutilibus ſubtilitatibus.

Cur alterius medio tabulæ percuſſa mini me frangantur.

HVIVSMODI ſunt, uelut cur ſi ſuper tabulam ictus inci dat, non adeò frangitur ſuppoſita tabula, ut in cuneis fi gendis; certè quia ictus per totam ſuppoſitã tabulam di ſtribuitur, unde minor euadit in ſingulis partibus; nec co gitur pars deſcendere alia relicta, quod maximè fracturę initium eſt, eo etiam intermedio temperatur ictus, aer quoq; cum impul ſu careat, uiolenter ingredi non poteſt ſecũdam tabulam ſubiectam, qua re nec diuidere.

Vitrea uaſa in igni expoſita ſurculo ſuppo ſito, cur non rumpantur.

Simile illud, cur uaſa uitrea igni expoſita ſuperimpoſito ſurculo non rumpantur: quoniam lignum non incaleſcit, cogitq; ut uitri partes mi ſceantur: frangitur aũt quia aqueum humidũ prius in flatũ conuertitur.

Inſtrumentum ludicrum.

Verum nullius uſus eſt inſtrumentum ex ſeptem annulis. Bractea fer rea digitum lata, palmi longitudine tenuis, in qua ſeptem foramina ro tunda, anguſta æquisq; ſpatijs, ſecundum longitudinẽ diſpoſita, ſeptem excipiunt uirgulas tenues, altitudine unciæ fermè, mobiles in imo, & in ſu prema parte circumflexas, ut annulos digiti magnitudine incluſos retine ant, ipſę uerò uirgæ à ſequenti annulo infra flexuram continentur. Ob idq; omnes annuli præter primum ab antecedente, ne exiliant liberè ex tra anteriorem uirgam prohibentur: ferrea omnia & ferrea etiam nauicu la, cuius ſpeciem ad unguem in margine reddidimus, longa lataq; pro magnitudine ſubiectæ laminæ. Hoc inſtrumento ludus excogitatus miræ ſubtilitatis. Primus ſecundusq; annulus per inane A ſpaciũ immittitur, inde nauicula per eoſdem annulos, poſt illorum primus per inane A de mittitur, poſt quem tertius annulus per nauiculæ uacuam mediam par tem, ut primi duo ſurſum trahitur; illiq; nauicula intruditur; tum etiam primo ſurſum deducto, iam tres circumambiunt nauiculam ipſam, demit tes igitur duos primos exempta prius nauicula, ita illa ſoli tertio inclu ſa manebit, inde quartum ſuperinducere licebit, ut omnis hæc induſtria tribus præceptis contineatur. Primum, quòd annulus ſurſum trahendus dimittendusue unum tantum habeat ante ſe cui nauicula includatur. Secundum, ut dum demittis unà ſemper primos duos demittas, & unum trahas uel unum demittendo duos primos trahas. Tertiũ, ut quocunq; ſurſum tracto uel demiſſo, omnes qui ante ſunt ſurſum trahere neceſſe eſt

Figure 8.16. Wallis [11, p. 473], and Cardan [2, p. 294].

[c]Some of this section was presented at the Fourth Gathering for Gardner, Atlanta, 2000.

Fourth Gathering for Gardner [Ch'ung En Yū. *Ingenious Ring Puzzle Book.* In Chinese: Shanghai Culture Publishing Co., Shanghai, 1958. English translation by Yenna Wu, published by Puzzles — Jerry Slocum, Beverly Hills, Calif., 1981]. On p. 6, it states that the puzzle was well known in the Sung Dynasty (960 1279). [There is a recent version, edited into simplified Chinese (with some English captions, etc.) by Lian Huan Jiu, with some commentary by Wei Zhang, giving the author's name as Yu Chong En, published by China Children's Publishing House, Beijing, 1999.]

8.8 Puzzle Grills

Traveling through medieval cities, one notices substantial window grills. Some of these had a puzzling central area where four rods formed a square with each rod passing through the next in sequence, see Figure 8.17. After some contemplation, one realizes that one could assemble such a square by a kind of uniform convergence. But the pattern continues outward and this prevents the uniform converging method. It makes a nice puzzle.

Escher uses the idea in his Cycle (1933) and Belvedere (1968), see Figure 8.18. However, examination shows Escher has continued alternating horizontal and vertical holes and the result appears to be genuinely impossible to assemble. Perhaps 20 years ago James Dalgety made a wire version which displayed the "easy" way to assemble/disassemble such a grill.

Other examples are given in Figure 8.19.

Figure 8.17. The Palazzo Thiene in Vicenza, designed by Palladio, with a close up of the same grill, showing the impossible central square.

Figure 8.18. Two uses by Escher, where both seem impossible.

Figure 8.19. From Città di Castello and the Bank of Italy, Florence.

After seeing my crude model[d] Simon Bexfield immediately programmed his 3D printer to produce a better model. In one day a good set of pieces was ready and we assembled them and showed it around. See Figure 8.20.

[d]I had given a short talk on these at "Maths Jam 2015" and had brought my own coat-hanger model.

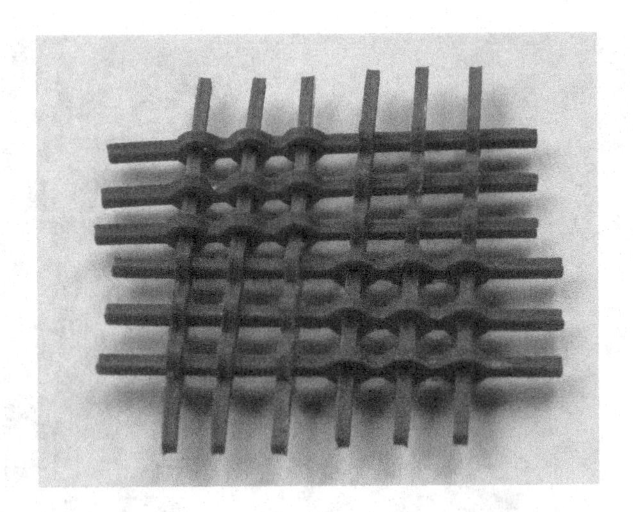

Figure 8.20. The first correct 3D printed example, made by Simon Bexfield, 7–8 November 2015.

8.9 Conclusions

Although recent work has pushed back our knowledge of these problems, sometimes by several hundred years, it seems clear that we really do not know the origins of them. None of the sources cited are massively original; they are simply compilations of well known examples of their time.

Bibliography

[1] CARDAN III.
[2] CARDAN IV.
[3] Leslie Daiken. *Children's Toys Throughout the Ages.* Spring Books, 1963.
[4] MAGICIAN'S OWN BOOK.
[5] MINGUÉT.
[6] OZANAM.
[7] PACIOLI-DVQ.
[8] PRÉVOST.
[9] SCHWENTER.
[10] J. Slocum and D. Gebhardt. *Puzzles from Catel's Cabinet and Bestelmeier's Magazine 1785 to 1823.* The Slocum Puzzle

Foundation, 1997. History of Puzzles Series. English translations of excerpts from the German Catel-Katalog and Bestelmeier-Katalog, with an introduction by David Singmaster.

[11] J. Wallis. *De Algebra Tractatus*, 1685, = *Opera Math.*, Oxford, 1693, vol. II, Chap. CXI, *De Complicatus Annulis*, 472–478.

[12] WECKER.

[13] WITGEEST-I.

[14] WITGEEST-II.

Interlude: Finding a Sardinian Maze

I enjoy traveling and looking for sites associated with mathematics. *The Art of the Maze*, by Adrian Fisher and Georg Gerster, describes and illustrates a maze carved on the wall of the Tomba del Labirinto in Luzzanas, Sardinia, dating from perhaps 2500–2000 BCE and this was thought to be the world's oldest example.[e]

My wife, Deborah, had taught in Florence in 1966 and made friends with an American couple of artists who had settled there. The wife, Ann Maury, is a distinguished botanical illustrator and was painting all the orchids of Italy, but she doesn't drive. There was one orchid still outstanding in Sardinia and this seemed like a good excuse for a holiday in June 1997 to Sardinia, where we had never been. In addition, I remembered the Luzzanas labyrinth, so thought we'd go look at it.

Sardinia is awash in archaeological sites: forts, tombs, sacred wells, menhirs, etc. When we asked the car agency where Luzzanas was, no one knew — it's not on any road map. We figured that we would stop at the archaeological museum in the center of the island, at Nuoro, and find out. Then we drove off over the mountains at the southeast corner of the island, which were quite spectacular — sheer rocks above and below the road.

We stopped at a village called Osini — this was new Osini, as old Osini had been abandoned due to the village sliding downhill. There we met the two Professors of Botany on Sardinia. We asked

[e]This was prepared for a family bulletin in 1998 and used on Adrian Fisher's web site and a shortened version was published — "The Oldest Labyrinth in Sardinia," in *Caerdroia: The Journal of Mazes & Labyrinths*, 30 (1999) 17–21.

both professors if they knew of Luzzanas or the labyrinth, but drew blanks. One, Professoressa Corrias, had a detailed road map and found a Luzzana near Sassari, but she actually lived near it and didn't think there was any tomb there. Both professors said they were friends with the archaeologists in their areas and would make inquiries for us.

Saturday morning, we packed up and went to the sea, then headed north along the most highly-rated drive on the island, which was up and down two wild valleys to Dorgali. The road is mostly at the top of the valley, with great views down into canyons. This was a main road, with the spectacular engineering associated with the Italian autostrade. Lots of tunnels and bridges, with a galleried section about a mile long. At one point, we came upon two families of wild pigs on the road. The area is mostly a national park and quite desolate.

After Dorgali, we turned inland and passed along a relatively flat area, with sheep, cattle and wild horses. We turned onto a more scenic route, passing under a large mountain called Sopramonte, looking like the high Sierras and apparently a refuge for bandits (hopefully in the past). Then into Nuoro in the late afternoon. Nuoro is one of the four provincial capitals on Sardinia. We followed signs to the Archaeological Museum, but found it had been shut for several years due to lack of staff. Professoressa Corrias had found the information and called us. It was near a town called Benetutti, about 12 miles from Nuoro, but we would have to get details there.

We arrived at Benetutti and stopped at the town hall. They said the Tomba was near some thermal baths (terme) on the road to Bultei and the staff there would be able to direct us and that it would be impossible to find without local help. We went along the road and came to the Terme Aurora which was closed up. A bit further on, beyond a small river, was the Terme San Saturnino, also closed. The area was fairly open countryside. We then drove around a bit, looking for farmhouses or dwellings or at least signs for the parking for the Tomba. At several points, we saw tracks going into buildings with a car, so we went along to them and found the cars were abandoned and no one was about.

Across from the Terme Aurora was an active farm with horses, but the gate to it was locked and no one seemed about. After a series of fruitless encounters, in great despair, we drove back to town and found it fairly shut up for siesta. We stopped at an open bar where we found one person who knew where it was — his class at school had done a project on it! After some more dead-ends we met up with the farmer's family. An animated discussion followed and the youngest boy said he knew where the Tomba was, but the others didn't seem to want him to say. We followed him back up the road about a mile, then into a field along a track, through a gate into another field and across it to another gate where he stopped and pointed to a pile of rocks about a quarter of a mile on. He said he had never gone into the cave as he was afraid of the frogs — we think his Italian wasn't too good and he probably meant bats. He left us there and we walked across the field of sparse grass about two feet tall, to the pile of rocks, where there was nothing. But the chap had said something like beyond the pile and the young man in the town had mentioned it was by the river, which was another fifty yards or so, where there was a mound rising a bit above the river bank. As we approached, we saw the mound was of stone and there were two round holes in it, about three feet across and filled with scraggly growth.

We walked around and over the mound, which dropped down about 15 feet to the stream. The upper floors of the Terme Aurora are clearly visible, about half a mile to the south. No other openings in the mound were apparent, so we examined the northern opening which had less dense shrubbery, mainly a small fig tree and some brambles. The hole was about three feet deep. I dropped into it and got out my trusty Swiss Army knife and cut away the growth and found an opening about two feet square cut into the rock at the bottom. I cleared away the growth around the opening and looked in to see a dry room. I had a small flashlight and crawled in to a room about eight feet square and three feet high with openings at the back and immediately to the left. And there, beyond the opening to the left, was the labyrinth carved on the stone!!! The others crawled in and sat beside me and then two bats flew about us. One bat stopped for a bit in the corner — they are really very sweet beasts.

At this point we wondered why the carving is attributed to perhaps 2500 BCE. Perhaps the style of the cave and some associated artifacts point to that date. But there was some modern graffiti on the same wall, so it's quite possible that the cave has been visited many times since it was made and the labyrinth could have been carved during the last century by some bored shepherd. (Jeff Saward writes that current archaeological belief is that the cave is Neolithic,

Figure 8.21. The author descends into the Tomba del Labirinto, Luzzanas, Sardinia, taken and provided by Ann Maury.

and the carving may be Roman.) The stone seemed to be limestone. It must not be too hard as one doesn't go carving chambers in really hard rock. It has been asserted that one can tell the nature of the tool that carved a pattern in rock — e.g. whether it was steel, iron, bronze, copper or bone. We don't know if such analysis has been applied here.

We left the cave, returned to our car and then drove off to visit other parts of the island before returning to London. This was certainly, for me, the most difficult to find mathematical site. Since then, Ann obtained the 1:25,000 map of the area of the Tomba del Labirinto. These are produced by the Istituto Geografico Militare and the relevant map is Foglio 481, Sezione III n Bono. (Bono is one of the towns near Benetutti.) The pile of stones is marked as an isolated building, though it has now collapsed. Written in small print across the fields between the road and the river is Luzzanas! However, this area does not seem to have any inhabitants! There is even a Nuraghe de Luzzanas about quarter of a mile to the NE of the mound. If the

Figure 8.22. The labyrinth graffito, provided by Jeff Saward.

pile of stones is the site marked on the map, then the universal map reference of the mound is 32TNKl0587621. With the map as a guide, one could walk NNW along the river by the Terme Aurora — the map shows another terme there, but it was a construction site when we were there. This is the Rio Mannu and one would go downstream (NNW) about 700 meters, to where the Mannu joins the Rio Tirso. Then one could go NE up the Tirso about 800 meters. However, we didn't see enough of this route to know whether it is passable. Even with the map, finding the Tomba will be difficult without some local assistance.[f]

[f] Jeff and Kimberley Saward were inspired to visit the Tomba — see "The Tomba del Labirinto Luzzanas, Sardinia." *Caerdroia*, 35 (2005) 5–11. This says the Tomba was first published in 1965 and says "To the best of our knowledge, the first written description of its location was given by David Singmaster in *Caerdroia* 30, subsequent to his successful visit in 1997, and without his notes and a detailed map of the area we would have struggled to find it." They think it is likely to date from the second half of the first millennium BCE. Jeff told me that he had tried to find the Tomba in 1973, but that was when J. Paul Getty Jr. had been kidnapped and was thought to be hidden on Sardinia, with the result that locals were reluctant to give any information!

Part II

New Ideas about Old Puzzles

Chapter 9

A Legacy of Camels

"Trust in God, but tie your camel"
—Arabic proverb.

The 17 Camels Problem is one of the best-known paradoxical problems in recreational mathematics. You may have heard some variation of it, but the telling with 17 camels is widespread. We give a version for the benefit of those who have not yet been perplexed by it.

> An Arab sheikh died leaving a herd of camels as his entire estate to be divided among his three sons. In his will, he specified that the eldest son should receive one half of the estate; the second son should receive one-third of the estate; and the third son should receive one-ninth of the estate. The sons went to examine the herd and found there were seventeen camels. (Sometimes it is stated that there had been eighteen but one had died.) Now seventeen is not divisible by two or three or nine and the sons were perplexed. Camels are valuable and they didn't want to cut one in pieces. After some discussion, they decided to consult the local Mullah Nasruddin and they sent for him. The Mullah rode over on his camel and listened to the sons' predicament. After some reflection, he said he would loan them his camel. The herd now had eighteen camels and the Mullah allotted half the camels to the eldest son, i.e. nine camels; then one-third of the camels to the second son, i.e. six camels; then one-ninth of the camels to the third son; i.e. two camels. Now there remained one camel, namely the camel that the Mullah had loaned them, so he reclaimed his camel and rode off into the sunset.

The key to the puzzle is that $\frac{1}{2} + \frac{1}{3} + \frac{1}{4}$ does not equal one, but equals $\frac{17}{18}$. The listener or reader does not normally notice this and is thoroughly mystified by the solution. We will look at generalizations of this with $\frac{1}{a} + \frac{1}{b} + \frac{1}{c} = 1 - \frac{1}{d}$ or $\frac{1}{a} + \frac{1}{b} + \frac{1}{c} + \frac{1}{d} = 1$. This can be adapted to the situation of two sons or four sons, etc. (In this chapter, all the denominators are positive integers.)

9.1 Some History

The given solution does not satisfy every one and there have been several authors who have claimed it is not correct and proposed alternative approaches. However, the problem is simply a relatively modern version of ancient problems of division where the parts do not add up to one. Such problems already occur in one of the oldest mathematical works, the *Rhind Papyrus* [24] of c. 1650 BCE. There, problem 63 wants to divide 700 loaves of bread in the proportion $\frac{2}{3} : \frac{1}{2} : \frac{1}{3} : \frac{1}{4}$. The total of the parts is $\frac{7}{4}$ and the ancient scribe says the unit is thus $\frac{4}{7}$ of 700, i.e. 400, so the first person gets $\frac{2}{3} \times 400 = 266\frac{2}{3}$ loaves. We would probably phrase this as the first person gets $\frac{2}{3}/\frac{7}{4}$ of 700.

Such problems with proportions that do not add up to one appear in many ancient sources, including several Indian works of the first millennium. By the middle ages, a more complex type of problem appeared. The first example seen is attributed to Gerhardt, in the Algorismus Ratisbonensis [14], where Problem 203 is to divide 384 into $\frac{2}{3}$ and 6 more, $\frac{3}{5}$ and 8 more, $\frac{5}{6}$ and 10 more, $\frac{7}{8}$ and 6 more. He takes a common denominator of 360 and finds $\frac{2}{3}$ of it is 240 and then adds 6 to get 246. Similarly, he gets 224, 310, 321 and then divides in the proportion 246:224:310:321. This problem is actually indeterminate as it depends on the choice of common denominator. The editor, Kurt Vogel, says the problem is unclear and the solution is false and notes that dividing 387 instead of 384 would give an integral solution.

Vogel cites a number of other occurrences of this problem, such as in Widman, Pacioli, Tonstall, Apianus, Cardan, Recorde and Buteo (1559), but it then seems to disappear, presumably because authors couldn't decide which approach was correct. Gerhardt [14], problems 170 and 286, are the earliest examples seen using $\frac{1}{2}$:$\frac{1}{3}$:$\frac{1}{9}$. Problem 286 discusses such problems, noting that $\frac{1}{2} + \frac{1}{3} + \frac{1}{6}$ leaves nothing, but $\frac{1}{2} + \frac{1}{3} + \frac{1}{9}$ leaves something, which appears to be the first comment about divisions where the parts do not add to one.

Later, Pacioli discusses $\frac{1}{2} + \frac{1}{3} + \frac{1}{4}$ and implies it is impossible because the fractions do not add to one, but then proceeds to divide in the usual proportion 6:4:3.

Tartaglia, in 1556 [30], gives a problem involving division into $\frac{1}{2} + \frac{1}{3} + \frac{1}{4}$. He has a long discussion of Luca Pacioli and others, marked "Error de frate Luca, & de molti altri" who assert that such problems are impossible or illegal. Tartaglia says simply to divide in the proportion 6:4:3 and cannot understand why others are making such a fuss.

Vera Sanford [25] says Tartaglia was among the first to suggest the use of the 18th camel—but does not say where in Tartaglia (the location is elusive). In an earlier work, Sanford [26] says that this is really a modern problem in that previously property was divided in proportion to fractions, regardless of whether they summed to unity. Tartaglia's [30] discussion of Pacioli [21] makes it clear that people were starting to object to this at this time, but examples continue and the modern version is not seen until late 19th century.

Pardon, in 1857 [22], asks to pay 20s with only 19s by dividing into $\frac{1}{2} + \frac{1}{3} + \frac{1}{6} + \frac{1}{19}$. "This, however, is only a payment upon paper." Note that the fractions add to $\frac{120}{114} = \frac{20}{19}$.

The first example of $\frac{1}{2} + \frac{1}{3} + \frac{1}{9}$ with animals, solved by loaning an animal seems to be in of 1872 [15], where it is called a Chinese puzzle and involves 17 elephants! The problem ends: "Query: Was the property divided according to the terms of the will?" but no further discussion occurs.

The problem with 17 horses is seen in 1881 and then in 1886, when Richard A. Proctor [23] gives it as "...the familiar puzzle [of] the farmer, ignorant of numbers, who left 17 horses to his three sons (or, equally well it may be, an Arab sheik who left 17 camels)." Proctor points out that if there were 35 camels, then the Cadi could also be left a camel. That is, we could divide 35 camels as $\frac{1}{2} + \frac{1}{3} + \frac{1}{9} + \frac{1}{36}$. Proctor's description may show the rapidity with which the problem circulated, or it may show that there are many other unseen occurrences of the problem. While these are not found in the significant puzzle books of this era, but puzzle problems of this sort were common in newspaper and periodical columns. For example, the next example is 17 elephants given in a humor book published in the UK and in Melbourne, Australia [9], which says the problem was in *The Galaxy* for August. *The Galaxy* may have been an Australian publication or an astronomical publication. Proctor was an astronomer who wrote popular science and had Australian connections, so this may refer to a version of his 1886 problem, though the form seems to be taken from *Hanky Panky* [15] which had a large circulation.

A 1888 version by Don Lemon [18] has 17 horses and a dervish. Hoffmann, 1893, [16] has a problem to divide as $\frac{1}{2}+\frac{1}{3}+\frac{1}{9}$. The answer says "this expedient is frequently employed" in "the Mahomedan Law of Inheritance" — probably a fictitious elaboration. A booklet of 1896 [8] has an early illustration of this puzzle, see Figure 9.1.

From about 1900 on, the problem regularly appears in puzzle books, but authors are often somewhat concerned about whether it is a correct solution and some give rather convoluted discussions. This is because they are unaware of the ancient problems of division which

AN UNMANAGEABLE LEGACY.

An old farmer left a will whereby he bequeathed his horses to his three sons, John, James and William, in the following proportions: John, the eldest, was to have one-half, James to have one-third, and William one-ninth. When he died, however, it was found that the number of horses in his stable was seventeen, a number which is divisible neither by two, by three or by nine. In their perplexity the three brothers consulted a clever lawyer, who hit on a scheme, whereby the intentions of the testator were carried out to the satisfaction of all parties. How was it managed?

Figure 9.1. *Brandreth Puzzle Book.* p. 3.

gave rise to this problem — the ancient problems disappeared from textbooks by about the mid-19th century. As seen, Tartaglia [30] had already dismissed such concerns.

- Sam Loyd, c. 1914 [19] discusses the problem and says the use of proportion makes the solution actually correct.
- Benson, 1904 [7], has "The lawyer's puzzle" which seems to be the first to say there were 18 horses, but one died.
- Ball/FitzPatrick, 1908–1909 [5], has a footnote stating the problem is Arabic.
- Ahrens, 1918 [2], cites Ball/FitzPatrick but says it has been in German oral tradition for a long time. He gives it with 17 horses.
- Dudeney, 1926 [12], has "The seventeen horses" and discusses the interpretation as a proportion, as in Ball/FitzPatrick, in detail.
- Kraitchik, 1930 [17], has 17 sheep, very similar to Ball–FitzPatrick, saying it is of Hindu origin.
- McKay in 1940 [20] has: "It is said that an Arab had 17 cattle."
- Zodiastar, 1941 [31] gives "The Arab and his steeds" dividing 19 horses into $\frac{1}{2} + \frac{1}{4} + \frac{1}{5}$, with a rather cryptic explanation of why the solution works.
- Philip E. Bath, 1959 [6] 7s (= 84 pence) divided $\frac{1}{2} + \frac{1}{4} + \frac{1}{8}$ by adding an extra shilling. Solution doesn't seem to understand this and claims it should be divided 42, 21, $10\frac{1}{2}$ with $10\frac{1}{2}$ left over.
- Doubleday, 1969 [11] has 17 cows divided $\frac{1}{2} + \frac{1}{3} + \frac{1}{9}$. He states the usual solution and then asks what is wrong with it. His solution notes that the fractions add to $\frac{17}{18}$ and then says "So, in making his will, the farmer hadn't distributed his entire herd." This seems confused to me as the entire herd has been distributed to the sons.
- D. B. Eperson, 1974 [13], gives The Shah's Rolls Royces, where one has to divide 23 Rolls Royces into $\frac{2}{3} + \frac{1}{6} + \frac{1}{8}$. The answer erroneously asserts this works for $n \equiv 1 (\text{mod } 24)$. What is true is that if $n = 23k$, then loaning k Rolls-Royces leads to the distribution $16k, 4k, 3k$ with k Rolls-Royces left over to repay the loan.
- My first "Brain Twister" for the *Weekend Telegraph* in 1988, [27], posed the problem for division of 41 oil wells as $\frac{1}{2} + \frac{1}{3} + \frac{1}{7}$, then asked for all quadruples like 2, 3, 7, 41.[a]

[a] I also suggested a version with 41 wives, but my editor chose the version with oil wells.

- John P. Ashley, 1997 [4], gives "Omar divides 17 horses among 3 sons." The answer says: "It was a great solution, but it was not correct mathematics. The sum of the fractional parts: $\frac{1}{2}, \frac{1}{3}$ and $\frac{1}{9}$ do not add up to 1 but to 17/18. Therefore, each of the heirs got a bit more than the will intended."

We have seen division of camels, money, elephants, horses, cows, Rolls-Royces, oil wells, wives. And the problem is claimed to be of Chinese, Islamic, Arabic, German traditional or Indian origin. Summarizing, the 17 camels problem is apparently a late 19th century version of an ancient problem, but it has spread through the world. There is no evidence of an Arabic, Hindu or Chinese origin, but this belief has certainly become widespread.

Pierre Ageron has an article on the history of this problem [1], which is not easy to verify. Imam 'Alî [ibn AbîTâlib] (601–664) was the cousin and son-in-law of the prophet Mohammed and was the designated successor. However, there was a dispute and a competitor became the first Caliph and 'Alî became Caliph in 656. 'Alî has become legendary for his ingenuity and many ideas are attributed to him, including the 17 camels problem. Unfortunately, there is no written documentation of this until Muhammad Mahdîal-Narâqî (1716–1795), an Iranian Shiite author in the 18th century. His work is in manuscripts, but includes *Mushkilât al-'ulûm* [The Problems of Sciences], which was printed in Teheran in 1988. Ageron gives a facsimile of the page, p. 90 of the 1988 version.

Ageron also describes some problems in this text. Several are inheritance problems. "Problem of the Pandects" — sharing a payment of 8 between those who contribute 5, 3 toward a meal. Find the least common multiple of 2, 3, ..., 10. He says there is an Arabic tradition associating the 17 Camels with Abraham ibn 'Ezra, but this is not borne out by the texts. He looks at several North African versions and numerous early European versions. He suggests the earliest oral version may be Yemeni, 15th century.

9.2 Analysis of the 17 Camels Problem

We have already seen that the key to the paradox is that $\frac{1}{2} + \frac{1}{3} + \frac{1}{9} = 1 - \frac{1}{18}$ and this leads us to ask for solutions of $\frac{1}{a} + \frac{1}{b} + \frac{1}{c} + \frac{1}{d} = 1$. More simply, we can look at the problem with two sons, which is looking for solutions of $\frac{1}{a}+\frac{1}{b}+\frac{1}{c}=1$. The problem with one heir is a bit unrealistic! In all these problems, the order of the values is not essential, so we will assume $a \le b \le c \le d \le \cdots$.

For the "two heirs problem," it is easy to find all solutions by hand. A computer program confirms our hand calculations. There are just three solutions for the two heirs problem.

No.	a	b	c
1	2	3	6
2	2	4	4
3	3	3	3

Each of these solutions does give a two heirs problem with $c-1$ camels, but none of these has ever been used! Perhaps it is too obvious that $\frac{1}{a}+\frac{1}{b}$ is not equal to one. Division in the proportion $\frac{1}{2}:\frac{1}{3}$ does occur a few times in the middle ages.

For the "three heirs problem" or

$$\frac{1}{a}+\frac{1}{b}+\frac{1}{c}+\frac{1}{d}=1,$$

it is not too hard to find all the solutions by hand, but it is a little tedious and one can make mistakes so they were also checked by computer. There are 14 solutions as follows.[b]

No.	a	b	c	d	Occurrences
1	2	3	7	42	Singmaster 1988
2	2	3	8	24	
3	2	3	9	18	about 20 occurrences
4*	2	3	10	15	
5	2	3	12	12	
6	2	4	5	20	6 occurrences
7	2	4	6	12	1 occurrence in 1897
8	2	4	8	8	Bath 1959
9	2	5	5	10	
10	2	6	6	6	

[b]I have a vague recollection of having seen this list at some time, but I have not been able to locate the source.

11	3	3	4	12
12	3	3	6	6
13*	3	4	4	6
14	4	4	4	4

Examining these solutions, we see that solutions 4 and 13 are unusual. In all other cases, we have that a, b and c divide d and we have a problem with $d-1$ camels divided as $1/a+1/b+1/c$ solved by loaning one camel. But in solution 4, we have $\frac{1}{2}+\frac{1}{3}+\frac{1}{10}=\frac{28}{30}$ which has been reduced to $\frac{14}{15}$ and 15 is not divisible by 10, so we cannot solve a 14 camels situation by loaning one camel. But we can solve a 28 camels problem by loaning two camels! Though requiring a slight extension of our perceptions, this seems close enough to our original situation that we call it a *pseudo-solution* and would want to find these as well as the original types of solutions. When we get to four heirs, we will examine the pseudo-solutions in more detail.

Of these solutions, only numbers 1, 3, 6, 7, 8 have occurred as inheritance problems. This is a bit surprising as puzzle setters usually explore most variations of a problem, and fairly quickly. Further, this problem is related to the Egyptian custom of expressing all fractions as the sum of unit fractions and it is a fairly natural question to ask how many ways one can be expressed as a sum of unit fractions[c]

The "four heirs problem" is obviously more extensive. My program found 97 solutions and 50 pseudo-solutions (i.e. requiring the loan of more than one camel). Surprisingly, none of these have ever been used as an inheritance problem! — though Proctor [23] mentions a situation with four heirs. The first solution corresponds to dividing 1805 camels as $\frac{1}{2}+\frac{1}{3}+\frac{1}{7}+\frac{1}{43}$. This is the unique case where $e=1806$ is the product of a,b,c,d and it has the largest value of e. The last solution is $a=b=c=d=e=5$, corresponding to dividing 4 camels as $\frac{1}{5}+\frac{1}{5}+\frac{1}{5}+\frac{1}{5}$.

Examining the pseudo-solutions leads to some interesting points. The first pseudo-solution has $a,b,c,d,e = 2,3,7,46,483$. We have $\frac{1}{2}+\frac{1}{3}+\frac{1}{7}+\frac{1}{46}=\frac{964}{966}$. Although this can be reduced to $\frac{482}{483}$, this cannot be used as a 482 camels problem because 483 is not divisible

[c]It seems I may have been the first to explicitly ask this question [28].

by a or d. But it can be viewed as a 964 camels problem solved by the loan of two camels.

Can we determine what is the minimal number of camels that occur in a pseudo-solution? We want $\frac{1}{a}+\frac{1}{b}+\frac{1}{c}+\frac{1}{d}=1-\frac{1}{e}=\frac{e-1}{e}$. Let L be the Least Common Multiple (LCM) of a,b,c,d. Then L/a is an integer and $1/a=(L/a)/L$, etc., so $\frac{1}{a}+\frac{1}{b}+\frac{1}{c}+\frac{1}{d}=\frac{\frac{L}{a}+\frac{L}{b}+\frac{L}{b}+\frac{L}{d}}{L}=\frac{e-1}{e}$. Denote the numerator $\frac{L}{a}+\frac{L}{b}+\frac{L}{b}+\frac{L}{d}$ as N, so $\frac{N}{L}=\frac{e-1}{e}$.

Now $\frac{e-1}{e}$ is in lowest terms, so we must have e divides L and $(e-1)$ divides N, indeed $N=L\frac{e-1}{e}$. When $e=L$ and hence $N=e-1$, we have an $e-1$ camels problem where only one camel is loaned. When $e<L$, we have an N camel problem where $L-N=L-(L-\frac{L}{e})=\frac{L}{e}$ camels have to be loaned.

However, it might be possible to reduce the situation by the common factor $G=GCD(\frac{L}{a},\frac{L}{b},\frac{L}{c},\frac{L}{d})$. But we have the following.

Lemma 9.1. *If L is the LCM of a set of numbers: $a,b,c,\ldots$, then the co-factors $\frac{L}{a},\frac{L}{b},\frac{L}{c},\ldots$, are relatively prime.*

Proof. Suppose g divides all of $\frac{L}{a},\frac{L}{b},\frac{L}{c},\ldots$, then $g|\frac{L}{a}$, so $ag|L$ and so $a|\frac{L}{g}$ and similarly $b|\frac{L}{g}$, $c|\frac{L}{g}$, ... and $\frac{L}{g}$ is a smaller common multiple of $a,b,c,\ldots$ than L is, contrary to the assumption that L is the least common multiple, unless $g=1$. □

Hence, in our problem, we cannot reduce the situation further and we see that there must be $\frac{L}{e}$ camels loaned.

Among the 50 pseudo-solutions are $a,b,c,d,e=2,3,7,78,91$ and 2, 3, 14, 15, 35, both of which correspond to inheritance problems where $\frac{L}{e}=6$ camels have to be loaned!

Note that $\frac{1}{3}+\frac{1}{5}+\frac{1}{6}+\frac{1}{8}=\frac{99}{120}=\frac{33}{40}$, so this does not fit into any of the forms seen. but could be a pseudo-problem for the division of 120 camels by the loan of 21 camels.

Some related problems are worth mentioning. Two students have to divide 47 items into $\frac{1}{3}+\frac{1}{4}+\frac{1}{5}$ [10]. The fractions add to 47/60, so they borrow 13. Dividing 60 into $\frac{1}{3}+\frac{1}{4}+\frac{1}{5}$ uses 47 items, leaving 13 to be returned. Though very similar to the 17 camels, this version works for any fractions, with the number of camels being the numerator of the sum of the fractions, or an appropriate multiple if necessary to make everything integral. For example, $\frac{1}{3}+\frac{1}{4}+\frac{2}{5}=\frac{59}{60}$ gives a problem where only one item has to be borrowed, while for $\frac{1}{3}+\frac{1}{4}+\frac{1}{6}=\frac{3}{4}$, we have to treat this as $\frac{9}{12}$ to get a pseudo-solution for 9 objects by borrowing 3.

There is a humorous version [29]. A teacher stood before his class and posed the following problem: "A wealthy man dies leaving an estate worth ten million pounds. One-third is to go to his wife, one-fifth is to go to his son, one-sixth to his butler, one-eighth to his secretary, and the rest to charity. Now, what does each get?" After a very long silence in the classroom, a hand was raised. The teacher called on the student. "A good lawyer?"

9.3 Analysis of the 13 Camels Problem

Apparently and surprisingly this problem has only appeared once, in Always, 1971 [3]. Always gives it with camels and an executor roughly as follows.

> An Arab sheikh died leaving a herd of camels as his entire estate to be divided among his three sons. In his will, he specified that the eldest son should receive one-half of the estate; the second son should receive one-third of the estate and the third son should receive one-quarter of the estate. The sons went to examine the herd and found there were thirteen camels. Now thirteen is not divisible by two or three or four and the sons were perplexed. Camels are valuable and they didn't want to cut one in pieces. After some discussion, they decided to consult the local Mullah Nasruddin and they sent for him. The Mullah rode over on his camel and listened to the sons' predicament. After some reflection, he borrowed a camel. The herd now had twelve camels and the Mullah allocated half the camels to the eldest son, i.e. six camels; then one-third of the camels to the second son, i.e. four camels. Now the third son's share is one-quarter of the herd which is three camels, but there are only two camels left. So the Mullah returned the camel he had borrowed and there were three camels to give to the third son. The Mullah then got on his camel and rode off into the sunset.

The solution of this problem might be considered somewhat less acceptable than in the previous cases, but the division is in the proportion 6:4:3 which is the ancient way to solve such problems and hence is just as valid as the previous cases. The key to the paradox is that $\frac{1}{2} + \frac{1}{3} + \frac{1}{4} = \frac{13}{12} > 1$.

So the generalization is to find other examples of $\frac{1}{a}+\frac{1}{b}+\frac{1}{c}=1+\frac{1}{e}$, etc. Again, we start with the simpler case of two heirs, which gives the equation

$$\frac{1}{a}+\frac{1}{b}-\frac{1}{c}=1.$$

The roles of a and b are interchangeable, so we will assume $1 \leq a \leq b \leq \cdots$. If $a=1$, then we must have $b=c$ and this gives an infinite family of solutions, though a rather trivial family. If $a=2$, then $b \geq 2$, so $\frac{1}{a}+\frac{1}{b} \leq 1$ and there is no solution. Recall that we are assuming that $a, b, \ldots$ are all positive integers.

Now consider the problem with three heirs, which gives us

$$\frac{1}{a}+\frac{1}{b}+\frac{1}{c}-\frac{1}{d}=1.$$

If $a \geq 3$, then $\frac{1}{a}+\frac{1}{b}+\frac{1}{c} \leq 1$ and there is no solution. If $a=2$ and $b=2$, then $c=d$, which gives another infinite, but fairly trivial, family of solutions. If $a=2$ and $b=3$, then $c \geq 6$ gives $\frac{1}{a}+\frac{1}{b}+\frac{1}{c} \leq 1$ and there is no solution. Hence we can only have $c=3,4,5$ and each of these has a solution, as follows.

No.	a	b	c	d	Occurrences
1	2	3	3	6	
2	2	3	4	12	Always 1971
3	2	3	5	30	

If $a=2$ and $b \geq 4$, then $\frac{1}{a}+\frac{1}{b}+\frac{1}{c} \leq 1$ and there are no solutions. The case $a=1$ remains, and this reduces immediately to

$$\frac{1}{b}+\frac{1}{c}=\frac{1}{d}. \tag{9.1}$$

An obvious, but again fairly trivial, family occurs with $b=c$, hence $b=c=2d$. We can eliminate this by assuming $b<c$. Consider any solution to Eq. (9.1). We can eliminate any common factor of b, c, d. Let $g=\mathrm{GCD}(b, c)$ and $\beta=\frac{b}{g}$, $\gamma=\frac{c}{g}$. It is well known and easy to show that $\mathrm{GCD}(\beta, \gamma)=1$. Then $b=\beta g$ and $c=\gamma g$. Setting these

into (9.1) gives us

$$\frac{\beta+\gamma}{\beta\gamma g} = \frac{1}{d}. \tag{9.2}$$

Lemma 9.2. $\mathrm{GCD}(\beta\gamma, \beta+\gamma) = 1$.

Proof. If not, then there is some prime p which divides both $\beta\gamma$ and $\beta+\gamma$. Then p must divide either β or γ. If p divides β, then p divides $\beta+\gamma$ implying p divides γ and so p is a common divisor of β and γ which have GCD of 1. Similarly for p divides γ. So there can be no such p, i.e. $\mathrm{GCD}(\beta\gamma,\ \beta+\gamma) = 1$. □

Since $\frac{1}{d}$ is in lowest terms, Eq. (9.2) gives us $\beta+\gamma = k$; $\beta\gamma g = kd = (\beta+\gamma)d$. And since $\mathrm{GCD}(\beta+\gamma, \beta\gamma) = 1$, we must have $\beta+\gamma$ divides g and $\beta\gamma$ divides d. So we can set $g = K(\beta+\gamma)$ and from $d(\beta+\gamma) = \beta\gamma g$ we have $d(\beta+\gamma) = K\beta\gamma(\beta+\gamma)$, so $d = K\beta\gamma$. But this also gives $b = \beta g = \beta K(\beta+\gamma)$ and $c = \gamma g = \gamma K(\beta+\gamma)$, so that K is a common factor of b, c and d, which we eliminated at the beginning. So we have to have $K = 1$ and all solutions of Eq. (9.1) with no common factor are given by $b = \beta(\beta+\gamma)$; $c = \gamma(\beta+\gamma)$; $d = \beta\gamma$, where β, γ are arbitrary, subject to $\mathrm{GCD}(\beta,\gamma) = 1$. The first few examples are as follows.

β	γ	$\beta+\gamma$	b	c	d
1	1	2	2	2	1

This leads to the first obvious group of solutions with $b = c$, so we assume $\beta < \gamma$ from here on.

1	2	3	3	6	2	
1	3	4	4	12	3	
1	γ	$\gamma+1$	$\gamma+1$	$\gamma(\gamma+1)$	γ	
2	3	5	10	15	6	
2	5	7	14	35	10	
2	γ	$\gamma+2$	$2(\gamma+2)$	$\gamma(\gamma+2)$	2γ	[Note that γ must be odd.]

3	4	7	21	28	12
3	5	8	24	40	15
		etc.			
4	5	9	36	45	20
		etc.			

However, we haven't checked that all these solutions can be interpreted as inheritance problems. The two heirs case works fairly trivially. For three sons, the cases with $a = b = 2$, $c = d$ depend on the parity. If c is odd, there is a $c + 1$ camels problem solvable by the Mullah borrowing one camel. But if c is even, there is a $2c + 2$ camels problem solvable by borrowing two camels. That is, this is a pseudo-solution. The three cases with $a = 2$, $b > 2$ are readily seen to yield such problems with $d + 1$ camels solved by the Imam borrowing one. The cases with $a = 1$, $b = c$ have a pseudo-solution with $b + 2$ camels solved by the Mullah borrowing two camels. After some experimentation, we find that the problem with parameters β, γ gives a pseudo-solution with $(\beta + \gamma)(\beta\gamma + 1)$ camels solvable by the Mullah borrowing $\beta + \gamma$ camels. For example, for the first interesting case: $\beta = 2$, $\gamma = 3$, we have the problem of dividing 35 camels as $\frac{1}{1} + \frac{1}{10} + \frac{1}{15}$. The Imam borrows 5 camels and then there are 30 camels. The first son gets all 30 and the other sons expect 3 and 2 camels, which are provided by the Imam returning the 5 camels he has borrowed.

The problem with four heirs gets a bit more involved and there are quite a number of trivial classes arising from the solutions for two or three heirs.

Case 1. If three of the reciprocals of a, b, c, d add to one, then equating the other to e gives solutions, e.g. a, b, c, d, e = 1, 2, 3, 6, 1; 2, 3, 3, 6, 3; 2, 3, 6, e, e; 3, 3, 3, e, e.

Case 2. We have a, b, c, d, $e = 1$, $3e$, $3e$, $3e$, e and 2, 2, $2e$, $2e$, e as solutions. More generally 1, b, c, d, e where $\frac{1}{b} + \frac{1}{c} + \frac{1}{d} = \frac{1}{e}$ is a solution, as is 2, 2, c, d, e where $\frac{1}{c} + \frac{1}{d} = \frac{1}{e}$. These cases can be eliminated by requiring $a > 1$, $e \neq a$, $e \neq b$, $e \neq c, e \neq d$ and prohibiting $a = b = 2$. In computing the solutions, one needs an upper bound for the values of d. If $\frac{1}{a} + \frac{1}{b} + \frac{1}{c} = x < 1$, we have $x + \frac{1}{d} = 1 + \frac{1}{e}$, so $\frac{1}{d} = (1 - x) + \frac{1}{e}$ which is a positive lower bound for $\frac{1}{d}$ which gives an upper bound for d. But when $\frac{1}{a}+\frac{1}{b}+\frac{1}{c}>1$, we can get arbitrarily large values of d in various of the trivial cases. After some computing and calculation, we find there are only a few of these situations to consider. As before, we assume $a \leq b \leq c \leq d$. The cases $a=1$ and $\frac{1}{a}+\frac{1}{b}=\frac{1}{c}=1$ have been considered already, so only $a=2$ can occur. The case $a=b=2$ has been considered. If $a=2$ and $b \geq 4$, then $\frac{1}{a}+\frac{1}{b}+\frac{1}{c}+\frac{1}{d}$ is at most one and there is no solution. So we must have $b=3$. Similar reasoning shows that c cannot be greater than 6, so we have just to examine $c=3,\ 4,\ 5$.

Case 3. $a=2, b=3, c=3$. Then we have $\frac{7}{6}+\frac{1}{d}=1+\frac{1}{e}$, so $\frac{1}{6}+\frac{1}{d}=\frac{1}{e}$. This gives $\frac{1}{d}=\frac{6-e}{6e}$, so we see that $e<6$ and there are only the following cases, some of which are already covered and some of which are impossible.

e	$6-e$	$6e$	d	a	b	c	d	e
1	5	6	–	–	–	–	–	–
2	4	12	3	2	3	3	3	2
3	3	18	6	2	3	3	6	3
4	2	24	12	2	3	3	12	4
5	1	30	30	2	3	3	30	5

Case 4. $a=2, b=3, c=4$. Then we have $\frac{13}{12}+\frac{1}{d}=1-\frac{1}{e}$ so $\frac{1}{d}=\frac{12-e}{12e}$ and $e<12$. This gives the following cases:

e	$12-e$	$12e$	d	a	b	c	d	e
3	9	36	4	2	3	4	4	3
4	8	48	6	2	3	4	6	4
6	6	72	12	2	3	4	12	6
8	4	96	24	2	3	4	24	8
9	3	108	36	2	3	4	36	9
10	2	120	60	2	3	4	60	10
11	1	132	132	2	3	4	132	11

Case 5. $a = 2, b = 3, c = 5$. Then we have $\frac{31}{30} + \frac{1}{d} = 1 - \frac{1}{e}$ so $\frac{1}{d} = \frac{30-e}{30e}$ and $e < 30$. This gives the following cases:

e	$30 - e$	$30e$	d	a	b	c	d	e
5	25	150	6	2	3	5	6	5
10	20	300	15	2	3	5	15	10
12	18	360	20	2	3	5	20	12
15	15	450	30	2	3	5	30	15
18	12	540	45	2	3	5	45	18
20	10	600	60	2	3	5	60	20
21	9	630	70	2	3	5	70	21
24	6	720	120	2	3	5	120	24
25	5	750	150	2	3	5	150	25
26	4	780	195	2	3	5	195	26
27	3	810	270	2	3	5	270	27
28	2	840	420	2	3	5	420	28
29	1	870	870	2	3	5	870	29

This covers all of the potentially unbounded cases and were among those found by the computer when d ranged up to 1000. In total there are 51 solutions and 32 pseudo-solutions. Let L be the LCM of a, b, c, d. Arguments similar to the 17 camels case shows that there is an $N + L/e$ camels problem where L/e camels have to be borrowed by the Mullah. Among these are the cases $a, b, c, d, e = 2, 3, 5, 195, 26$ and 2, 3, 5, 420, 28 where 15 camels have to be borrowed and the case 2, 3, 5, 870, 29 where the Mullah has to borrow 30 camels!

Bibliography

[1] P. Ageron. "Le partage des dix-sept chameaux et autre exploits arithmétiques attribués à l'imam 'Alî Mouvance et circulation de récits de la tradition musulmane chiite." *Revue d'Histoire des Mathématiques*, 19, 1 (2013) 1–41.

[2] W. Ahrens. *Altes und Neues aus der Unterhaltungsmathematik.* Springer, 1918, 84–85.

[3] J. Always. *Puzzles for Puzzlers.* Tandem, 1971. Prob. 35: An odd bequest, 22 & 71.

[4] J. P. Ashley. *Arithmetickle.* Keystone Agencies, 1997, 8.

[5] W. W. R. Ball. *Récréations et Problmes Mathématiques des Temps Anciens & Modernes.* Hermann, Paris. French translation by J. Fitz Patrick of BALL. (1st ed., from the 3rd ed, 1896, of BALL, "Revue et augmentée par l'auteur"; 1898. 2nd ed., from the 4th ed, 1905, of BALL, "et enrichie de nombreuses additions", 1908/1909. Part 1, p. 111. The problem is not in the 1st ed., nor in Ball, 4th ed.) The 2nd ed. has been reprinted in both three volumes and one large volume.

[6] P. E. Bath. *Fun with Figures.* The Epworth Press, 1959. No. 66: "The vicar's garden," 25 & 52.

[7] J. K. Benson. *The Book of Indoor Games for Young People of All Ages.* C. Arthur Pearson, 1904. "The lawyer's puzzle," 225. Also in J. K. Benson, ed. *The Pearson Puzzle Book.* C. Arthur Pearson, c. 1910, 52.[d]

[8] *Brandreth Puzzle Book.* Brandreth's Pills, 1896.

[9] E. W. Cole [or perhaps Arthur C. Cole]. *Cole's Fun Doctor: The Funniest Book in the World.* Routledge, London & E. W. Cole, Melbourne, no date [1886] 224.

[10] C. Cook. *Mathematical Pie*, 23 (February 1958) 178.

[11] E. Doubleday. *Test Your Wits* Vol. 1. Ace Publishing, 1969, Prob. 25 "Milk shake," pp. 36 & 159.

[12] Henry Ernest Dudeney. *Modern Puzzles.* C. Arthur Pearson, 1926 (also new ed. circa 1936) Prob. 89 "The seventeen horses," 33–34 & 123–124. (Included in *536 Puzzles and Curious Problems.* Ed. by Martin Gardner. Scribner's, 1967, prob. 172, pp. 54–55 & 266–267. also Fontana, 1970, vol. 1, prob. 84, 34 & 125.

[13] D. B. Eperson. "Puzzles, pastimes and problems." *Mathematics in School,* 3:6 (November 1974) 12–13 & 26–27. Prob. 6 "The Shah's Rolls Royces".

[14] GERHARDT.

[15] W. H. Cremer, Jr. (Ed.) *Hanky Panky: A Book of Easy and Difficult Conjuring Tricks.* John Camden Hotten, 1872. Also several editions c. 1873 from Chatto & Windus. My edition is c. 1875 from John Grant. This is one of six similar books often attributed to Wiljalba Frikell which copied from one another. In fact some bibliographers think this one is by Henry Llewellyn Williams.

[16] Professor Louis Hoffmann [pseudonym of Angelo John Lewis]. *Puzzles Old and New.* Warne, 1893. (Also Martin Breese, 1988 and Hordern,

[d]There is a much different book with the same title and cover, but edited by Mr. "X"; C. Arthur Pearson, London, (1921), 2nd ed., 1923.

1993, p. 119.) Chap. IV, no. 11, "An Unmanageable Legacy", 147 & 191–192.

[17] M. Kraitchik. *La Mathématique des Jeux ou Récréations Mathématiques.* Stevens, 1930, Chap. 1, prob. 47, p. 15. (The material is not in the 2nd edition or the English translation.).

[18] D. Lemon. *Everybody's Pocket Cyclopedia* .... Saxon & Co., (1888) (also revised 8th ed., 1890.) No. 2, 135.

[19] S. Loyd. "Problem 37: A queer legacy." *Tit Bits,* 32 (5 June and 3 July 1897) 173 & 258. (also LOYD. "The herd of camels," 57 & 346.)

[20] H. McKay. *Party Night.* OUP, 1940. "Arithmetical Catches and Puzzles," no. 35, 184.

[21] PACIOLI-SUMMA.

[22] Uncle George [George Frederick Pardon]. *Parlour Pastime for the Young.* James Blackwood, 1857. "Arithmetical puzzles," no. 4, p. 173. (The later edition, *Parlour Pastimes,* has our problem on p. 184.)

[23] R. A. Proctor. "Some puzzles." *Knowledge,* 9 (August 1886) 305–306.

[24] THE RHIND PAPYRUS. Problem 63. Chace, 101 of vol. 1. Robins & Shute, p. 41 & plate 18.

[25] V. Sanford. *A Short History of Mathematics.* Houghton Mifflin, 1930 & 1958. 218–219.

[26] V. Sanford. *The History and Significance of Certain Standard Problems in Algebra.* Teachers College, Columbia University, *Contributions to Education,* No. 251, 1927, 87.

[27] D. Singmaster. "A Middle Eastern muddle." *Games & Puzzles,* 12 (March 1995) 18–19 & 13 (April 1995) 40. (First appeared as "Well, well, well." in *Weekend Telegraph* (27 February 1988, 5 March 1988 and 12 March 1988).)

[28] D. Singmaster. "The number of representations of one as a sum of unit fractions."[e] in R. K. Guy. *Unsolved Problems in Number Theory.* Springer, 1981. Prob. D11, 89 (also 2nd ed, 1994, 158–166).

[29] J. Swan. *Another Man Walks Into A Bar,* Ebury Press, 2007, p. 141.

[30] TARTAGLIA. Part 1, Book 12, art. 42, ff. 200r–200v.

[31] "Zodiastar." *Fun with Matches and Matchboxes Puzzles, Games, Tricks, Stunts, Etc.* Universal, circa 1941, 54-56 & 75.[f]

[e] Basic work done in 1972–1973.

[f] There are many variants but generally identified as UPL Book 44.

Chapter 10

Heronian Triangles

Most readers will know about Pythagorean triples. These are integers x, y, z, such that

$$x^2 + y^2 = z^2 \tag{10.1}$$

and the motivation is the famous Pythagorean theorem for right triangles with sides of lengths x, y, z. Knowledge of these goes back to Old Babylonian times — the cuneiform tablet Plimpton 322 from c. 1800 BCE gives a table of such triples which follows a rule described in Euclid X.28, c. 300 BCE.

However, there is another type of ancient triangle, named for Hero (or Heron) of Alexandria, c. 150 CE. Like right triangles, they are characterized by the lengths of their sides, but it is their relationship to the area of the triangle Δ that matters. Heron gave us a formula that says if $s = (a + b + c)/2$ then $\Delta^2 = s(s - a)(s - b)(s - c)$, (where s is called the semi-perimeter). For example, for $a, b, c = 13, 14, 15$, then $s = 21$, $\Delta^2 = 7056$, $\Delta = 84$. Heron's formula was known to Archimedes, c. 250 BCE, and to the Indian mathematician Brahmagupta, c. 620.

A *Heronian* triangle is a triangle with sides a, b, c and area, Δ, all integers. In other words, when not only are the sides integers but the area is an integer too. Despite the ancient knowledge of the formula and much work (Dickson [2] has a whole chapter on the topic), the problem of determining all Heronian triangles was not accomplished until 1979!

10.1 Determination of Pythagorean triples

Because the determination of Heronian triangles is similar to the determination of Pythagorean triangles and uses some of the same ideas and results, we first describe the determination of Pythagorean triples, which can be found in most introductory books on number theory.

A Pythagorean triple x, y, z is *primitive* if each pair of sides has no common factor. From (10.1) it follows that any common prime factor of two sides is also a prime factor of the third side. For any Pythagorean triple, x, y, z, we can divide all the sides by any common prime factor to get an associated primitive triple. For example, 6, 8, 10 is a Pythagorean triple whose associated primitive triple is 3, 4, 5. It follows that at most one side of a primitive triple can be even. If x and y are both odd, say $x = 2a + 1, y = 2b + 1$, then $x^2 = 4(a^2 + a) + 1$ and $y^2 = 4(b^2 + b) + 1$ so both are congruent to $1 (\mathrm{mod}\, 4)$, which makes z^2 congruent to $2 \,(\mathrm{mod}\, 4)$, which is easily seen to never hold, since an even square must be divisible by 4. So one of x, y must be even and the other must be odd, and z is odd. Let us take y as even, $y = 2w$. Then $y^2 = 4w^2 = z^2 - x^2 = (z + x)(z - x)$. Since both $z + x$ and $z - x$ are even, we can write $w^2 = (z + x)/2 \cdot (z - x)/2$. Since any prime factor of $(z+x)/2$ and $(z-x)/2$ would divide their sum z and their difference x, it follows that these two factors must be coprime. From the fact that integers enjoy unique factorization, it follows that these factors must themselves be squares, say $p^2 = (z + x)/2$ and $q^2 = (z - x)/2$, where $p > q \geq 1$. Then $z = p^2 + q^2$, $x = p^2 - q^2$, and $y = 2pq$. For primitiveness, it is necessary and sufficient that p and q are coprime and not both odd. Here are the first examples.

p	q	x	y	z
2	1	3	4	5
3	2	5	12	13
4	1	15	8	17
4	3	7	24	25
5	2	21	20	29
5	4	9	40	41

10.2 Determination of Heronian triples

As already noted, there have been many attempts to do this. Carlson's article [1] contained some oversights which were corrected in [9]. This seems to be the first complete determination of Heronian triangles. Recently Posamentier *et al.* [7] discussed Heronian triangles without giving this result.[a]

Much of this discussion comes from L. E. Dickson [2]. Unfortunately, he uses the term Heronian for triangles with rational sides and area. We shall refer to these as rational, reserving Heronian for triangles with integral sides and area. It is clear that knowledge of either type of triangle determines the other type.

The general idea of juxtaposing Pythagorean triangles to get a rational triangle seems to be due to G. C. Bachet in his *Commentary on Diophantus* (1621) [2] (item 3). Euler [2] (item 7) was the first to give a formula for all rational triangles. Unfortunately, this is contained in posthumous, undated, papers and the proof is lacking. D. N. Lehmer [2] (item 43) appears to be the first to derive Euler's results and to publish a determination of all rational triangles in 1899–1900. It is quite easy to derive Euler's results, so there seems little doubt that Euler himself had done so. However, without detailed examination of sources, it is difficult to determine priority, especially since Dickson's notes are often rather condensed. For instance, Brahmagupta (7th century) [2] (item 1) gives a formula which is the same as that obtained by a juxtaposition (but it is not clear if he did it this way or treated it as a general process) and which is essentially equivalent to Euler's formula. There are also several papers after Euler and before Lehmer which use juxtaposition and/or Euler's formula, but Dickson's notes are not clear as to whether they show the completeness of the formula. Lehmer's derivation uses the rationality of the sines and cosines of the angles.

Dickson noted that the Japanese mathematician Nakane Genkai first found all the Heronian triangles with consecutive integer sides in 1722 [2] (item 6a), though Dickson could not tell if the proof was complete. W. Sierpinski [8] proves that "each rational triangle can

[a] This section originally appeared as a letter to the editor [10].

be obtained by the union of two right-angled triangles with rational sides" and he observes that not every Heronian triangle is obtained as a union of two Pythagorean ones. He also finds the Heronian triangles with consecutive integer sides.

A. P. Domoryad [3] says "it is easy to prove (21) that any of the altitudes of a 'Heronic triangle' ... gives two right-angled triangles with rational sides ..." Note "(21)" is explained on p. 253, though it is rather cryptic due to three misprints in one sentence.

Oystein Ore [5] says "we have no general formula giving them all," referring to Heronian triangles. John R. Carlson [1] built on Ore's work; see also [9]. In addition Carlson makes the observation that, if the sides of a Heronian triangle have a common factor, then, when the sides are divided by this factor, the resulting triangle is also Heronian. However, Carlson's short proof (and Pargeter's longer proof [6]), both are incomplete. Carlson accidentally assumes that $s = (a + b + c)/2$ is an integer, while Pargeter shows that a, b, c cannot all be odd and then asserts that they must all be even. But his proof actually shows that $t = 2s$ can not be odd, which is what is needed. The following combines features of both proofs.

Theorem. *Let k, a, b, c be positive integers. Then a, b, c are the sides of a Heronian triangle if and only if ka, kb, kc form a Heronian triangle.*

Proof. The "only if" is clear. Suppose ka, kb, kc are the sides of a Heronian triangle of area B. Let $t = a + b + c$. If t is even, then the semi-perimeter $s = t/2$ is an integer and the area Δ of the triangle with sides a, b, c is $\Delta = \sqrt{s(s-a)(s-b)(s-c)}$ which is the square root of an integer. Now B is an integer and $B = k^2\Delta$, so Δ is rational. But a rational square root of an integer must be an integer, so we are done when t is even.

Pargeter's argument carries through almost verbatim to show that t cannot be odd. For the sake of completeness it is reproduced here in condensed form. Suppose t is odd and let $u = t - 2a$, $v = t - 2b$, $w = t - 2c$, so $t = u + v + w$ and u, v, w are all odd. Now $16B^2 = 16k^4\Delta^2 = k^4 tuvw$ is an integral square. Considering congruence classes $(\text{mod}\, 4)$ for u, v, w, one sees that $tuvw \equiv 3 \,(\text{mod}\, 4)$ always holds. But an odd square is always $\equiv 1(\text{mod} 4)$. Thus t cannot be odd. □

By re-parametrizing; our result can simplified, as follows: A triangle is Heronian if and only if its sides can be represented as:

(i) $a(u^2+v^2)$, $b(r^2+s^2)$, $a(u^2-v^2)+b(r^2-s^2)$, where $auv = brs$, or

(ii) a reduction by a common factor of a triangle given by (i).

Because of (ii), there is no harm in multiplying (i) through by $uvrs$ and then canceling $auv = brs$. This gives Euler's formula again.

Posamentier *et al.* [7] discuss Heronian triangles, but are not aware of this result. They give 5, 29, 30 as a Heronian triangle which is not obtained by juxtaposition. Setting $a = 1$, $u = 4$, $v = 3$, $b = 1$, $r = 12$, $s = 1$ in (i) gives 25, 145, 150 and this reduces by the common factor 5 to give 5, 29, 30.

Pargeter says "it is much easier to make Heronian triangles by fitting together Pythagorean triangles with equal sides (adjacent to the right angles). Such triangles, reduced if necessary, will always have one integral altitude." More recent work can be found in [4].

Bibliography

[1] J. R. Carlson. "Determination of Heronian Triangles." *Fibonacci Quarterly,* 8 (1970) 499–506 & 551.

[2] L. E. Dickson. "Triangles, Quadrilaterals and Tetrahedra with Rational Sides: Rational or Heron Triangles." In *History of The Theory of Numbers,* Volume II: Diophantine Analysis. G. E. Stechert & Co, 1934 (also Chelsea Publ. 1952) Chapter V: 191–201.

[3] A. P. Domoryad. *Mathematical Games and Pastimes.* Pergamon, Oxford, 1963 (Russian original, 1961), 31–32.

[4] W. J. LeVeque, ed. *Reviews in Number Theory.* American Mathematical Society, Providence, RI, 1974, Vol. 2, p. 72.
For convenience of those who do not have access to this collection, the items are from *Mathematical Reviews*: 22 (4656); 23 (A107); 27 (4792); 31 (121).

[5] O. Ore. *Invitation to Number Theory.* Random House, 1967, 59–60.

[6] A. R. Pargeter. "Mathematical Notes 2895(2): Heronian triangles." *Mathematical Gazette.* 44, 348 (May 1960) 130–131.

[7] A. S. Posamentier, R. Geretschläger, C. Li and C. Spreitzer. *The Joy of Mathematics.* Prometheus Books, 2017, 127–131.

[8] W. Sierpinski. *Pythagorean Triangles.* Yeshiva University, 1962, pp. 59–66.

[9] D. Singmaster. "Some Corrections to Carlson's 'Determination of Heronian Triangles'." *Fibonacci Quarterly*, 11, 2 (April 1973) 157–158.

[10] D. Singmaster. "Letter: Heronian triangles." *Mathematical Spectrum* 11, 2 (1978/79) 58–59.

Chapter 11

The Ass and Mule Problem

> The mule says to the ass: "If you give me one of your sacks, I will have as many as you." The ass responds: "If you give me one of your sacks, I will have twice as many as you." How many sacks did they have?

The general version of the problem (for two individuals) is the situation where the first says: "If I had a from you, I'd have b times you," and the second responds: "And if I had c from you, I'd have d times you." It is traditional for the parameters a, b, c, d and the solutions, say x, y, to be integers. The parameters will be written $(a, b; c, d)$.

The first known Ass and Mule Problem is attributed to Euclid — see [3] for (1, 2; 1, 1) in Greek and Latin verse. The general problem is treated by Diophantus [1] and he does (30, 2; 50, 3) as an example. Versions appear in The Greek Anthology (see Chapter 2) and Bhāskara (Chapter 3) and Alcuin (Chapter 4).[a]

It has been attributed to Fibonacci and this is true for the case (7, 5; 5, 7). Fibonacci, as usual, studies the problem in great detail, with many variants, taking about 14 pages [4], and he later goes over the problems in his material on false position. For example, he considers the extended problem where the first says "If I had 7 from you, I'd have 5 times you plus 1 more", i.e. the first equation is $x + 7 = 5(y - 7) + 1$. However, Fibonacci is not the first to

[a]This chapter appeared in *Mathematical Gazette* 89 (No. 516) (November 2005) 365–370.

have non-integral solutions. The two problems in *The Greek Anthology* are $(10,3;10,5)$ and $(2,2;2,4)$ whose solutions both come out in sevenths, see Chapter 2. Fibonacci is apparently the first to use non-integral parameters — he does $(6,5\frac{1}{4};4,7\frac{2}{3})$. In fact, Greek and medieval mathematicians had no difficulties in dealing with complex fractions. Fibonacci has similar problems with denominators 50, 61, 721 and 394.

11.1 Analysis of the Original Problem

Is there a simple way to know when integral parameters will yield integral solutions? Unlike Fibonacci, we will only be concerned with integers. The question of when integral data gives integral solutions was difficult. My result [5] is given here.

The problem with parameters $(a,b;c,d)$ leads to these equations:

$$x+a=b(y-a); \quad y+c=d(x-c). \tag{11.1}$$

The solutions are readily computed as:

$$x=c+\frac{(b+1)(a+c)}{bd-1}; \quad y=a+\frac{(d+1)(a+c)}{bd-1}. \tag{11.2}$$

Note that this system of equations implies that x is integral if and only y is integral.

The values of x and y are integers if and only if $bd-1$ divides both of $(b+1)(a+c)$ and $(d+1)(a+c)$, which is if and only if $(bd-1)$ divides $\mathrm{GCD}((b+1)(a+c),(d+1)(a+c)) = (a+c)\,\mathrm{GCD}(b+1,d+1)$.(GCD is the greatest common divisor function.) Now consider $g=\mathrm{GCD}(b+1,d+1)$. This g must divide $(b+1)(d+1)-(b+1)-(d+1)=bd-1$. Hence, the last statement of the previous paragraph gives us that x and y are integers if and only if

$$\frac{bd-1}{\mathrm{GCD}(b+1,d+1)} \text{ divides } a+c. \tag{11.3}$$

This seems to be as simple a criterion for integrality as one could hope for. This criterion allows us to pick arbitrary b and d, assuming $bd\neq 1$, and then determine which values of a and c give integral solutions. This does not depend on the signs of any of the parameters. It is particularly striking that a and c only enter via the sum $a+c$.

11.2 Došlić's Variation

Došlić considers the problem $(a, b; b, a)$. This variant was solved [2] but our new and more general analysis builds on the solution just given for the classic problem. So by setting $c = a, d = b$ the problem now is:

$$x + a = b(y - a); \quad y + b = a(x - b), \tag{11.4}$$

with solutions:

$$x = b + \frac{(b+1)(a+b)}{ab-1}; \quad y = a + \frac{(a+1)(a+b)}{ab-1}. \tag{11.5}$$

The condition for integrality becomes:

$$\frac{ab-1}{\text{GCD}(a+1, b+1)} \text{ divides } (a+b). \tag{11.6}$$

Surprisingly, the integral solutions of this special case are harder to determine than for the general case.

The problem is symmetric in a and b in that interchanging a and b also interchanges x and y. So we can assume $a \leq b$. Došlić assumes a and b are positive, so the case $ab = 1$ is the same as $a = b = 1$ and this has no solution. (Later we relax this assumption.) So we now assume $ab > 1$.

The second equation of Eq. (11.5) shows that y is integral if and only if $(ab - 1)$ divides $(a + 1)(a + b)$, and our earlier remark shows this is equivalent to x being integral. Subtracting $(ab-1)$ from $(a + 1)(a + b)$, this is equivalent to:

$$\frac{1 + a + a^2 + b}{ab-1} \text{ is an integer.} \tag{11.7}$$

This expression is 1 if and only if $2 + a + a^2 = b(a - 1)$. The case $a = 1$ cannot occur here. So the last equation holds if and only if $b = (2 + a + a^2)/(a - 1) = a + 2 + 4/(a - 1)$ and for this to be an integer, we must have $a = 2, 3, 5$ with corresponding $b = 8, 7, 8$ and these are solutions of the problem.

If the expression in Eq. (11.7) is an integer but not one, then it must be at least two, and this leads to $3 + a + a^2 \geq (2a - 1)b$. This leads to $0 \geq a^2 - 2a - 3 = (a - 3)(a + 1)$, recalling that we want $b \geq a$. For positive integral a, this requires $a = 1, 2, 3$, for which $b \leq 5, 3, 3$ respectively. So the only cases to consider are $a, b = 1, 2;\ 1, 3;\ 1, 4;\ 1, 5;\ 2, 2;\ 2, 3;\ 3, 3$. All of these, except $1, 4$, give solutions and so we have precisely the same 16 solutions that Došlić found, namely: $1, 2;\ 1, 3;\ 1, 5;\ 2, 2;\ 2, 3;\ 2, 8;\ 3, 3;\ 3, 7;\ 5, 8$ and the reversals of seven of these.

Non-positive Solutions

What about $ab = 1$? Note $a = b = -1$ was feasible and indeed has solutions. This led to the realization that one can also find all the non-positive solutions.

First, we dispose of three special cases.

- If $ab = 1$, we have $a = b = \pm 1$. The subcase $a = b = 1$ leads to an impossible equation. The subcase $a = b = -1$ leads to $y = -x$, for any x.
- If $ab = 0$, we have $a = 0$ or $b = 0$. Recalling $a \leq b$, we see there are three subcases: $a = b = 0$; $a = 0,\ b > 0$; $a < 0,\ b = 0$. Each of these has solutions: $x = y = 0$; $x = -b^2,\ y = -b$; $x = -a,\ y = -a^2$ respectively.
- If $a = -1$, the second part of Eq. (11.4) gives $y + b = -x + b$, so $y = -x$. Then the first of part of Eq. (11.4) gives $x - 1 = -b(x - 1)$ so $x = 1,\ y = -1$ is a solution for any b, or we have $b = -1$, which case was considered above. The case $b = -1$ behaves similarly.

The previous section has found all solutions with positive a, b. After noting these special cases, we have two cases remaining to consider: (1) $a \leq b < -1$, (2) $a < -1, 1 \leq b$.

The analysis in the previous section works with $(a + 1)(a + b)/(ab - 1)$. But the assumption $a \leq b$ makes it easier to now work with $(b + 1)(a + b)/(ab - 1) = 1 + B$, where $B = (1 + b + b^2 + a)/(ab - 1)$. Note x and y are integral if and only if B is integral. There are two cases.

- $a \leq b < -1$. In this case $b+1 < 0$; $a+b < 0$; $ab > 1$; $ab-1 > 0$, so $1+B > 0$ and $B \geq 0$ (since B is integral). Now we show that $B \geq 1$ cannot hold.

  We have $B \geq 1$ if and only if $1+b+b^2+a \geq ab-1$. A little rearrangement shows this is if and only if $(b+1)^2 \geq (a+1)(b-1)$. Now we have $a+1 \leq b+1 < 0$ and $b-1 < b+1 < 0$, so $(a+1)(b-1) \geq (b+1)(b-1)$ and $(b-1)(b+1) > (b+1)^2$, hence $(a+1)(b+1) > (b+1)^2$, which shows that $B < 1$. So if we have an integer solution, we must have $B = 0$ and $1+B = 1$. The first part of Eq. (11.5) then gives us $x = b+1$, and this gives us $a = -(b^2+b+1), y = -(b+1)^2$, for any b.
- $a < -1, 1 \leq b$. Now it is not so obvious that $B \geq 0$. But $B < 0$ is equivalent to $B \leq -1$ which is if and only if $1+b+b^2+a \leq ab-1$. Rearrangement shows this is equivalent to $b(b+1) \leq a(b-1)$. But $a < b, 0 < b$ and $0 \leq b-1 < b+1$ gives us $a(b-1) \leq b(b-1) < b(b+1)$, so $B \leq -1$ is impossible and we have $B \geq 0$.

  Now we show that $B \geq 1$ cannot hold. We have $B \geq 1$ if and only if $1+b+b^2+a \leq ab-1$. A little rearrangement shows this is if and only if $(b+1)^2 \leq (a+1)(b-1)$. Now we have $a+1 < 0 < b+1$ and $0 \leq b-1 < b+1$, so $(a+1)(b-1) < (b+1)(b-1)$ and $(b-1)(b+1) < (b+1)^2$, hence $(a+1)(b+1) < (b+1)^2$, which shows that $B < 1$. So if we have an integer solution, we must have $B = 0$ and $1+B = 1$. The first part of Eq. (11.5) then gives us $x = b+1$, and this gives us $a = -(b^2+b+1), y = (b+1)^2$, for any b.

So we see that, beyond the positive cases and the special cases treated at the beginning of this section, there is a solution for each b.

11.3 Another Simpler Variation

> Alice and Bob were comparing their stacks of pennies. Alice said "If you gave me a certain number of pennies from your stack, then I'd have six times as many as you, but if I gave you that number, you'd have one-third as many as me." What is the smallest number of pennies that Alice could have had?

The general form of this new variation is the situation where the first says: "If I had a from you, I'd have b times you, but if I gave c to you, I'd have d times you." This leads to the equations:

$$x + a = b(y - a); \quad d(y + c) = x - c. \tag{11.8}$$

The integral d of the classic problem has been changed to $1/d$ with d integral. Note that $b > d$ for reasonable solutions. In the particular problem given, we have $b = 6$, $d = 3$ and $a = c$ is an unknown, but our analysis turns out to deal with this easily.[b]

We can solve Eq. (11.8) directly or use Eq. (11.2), obtaining

$$x = c + \frac{d(b+1)(a+c)}{b-d}; \quad y = a + \frac{(d+1)(a+c)}{b-d}. \tag{11.9}$$

One can see from Eq. (11.9), and it is obvious from Eq. (11.8), that x is an integer if y is an integer, but the converse does not hold (consider $c = 1, d = 2, x = 4$). The value of y is an integer if and only if $(b-d)$ divides $(d+1)(a+c)$, which is if and only if $\text{GCD}(b-d, d+1)$ divides $a + c$. Now consider $g = \text{GCD}(b - d, d + 1)$. This is equal to $\text{GCD}(b+1, d+1)$. Hence, as earlier, x and y are integers if and only if

$$\frac{b-d}{\text{GCD}(b+1, d+1)} \text{ divides } (a+c). \tag{11.10}$$

(If these statements are not obvious, proving them gives a pleasant exercise.)

Again, this seems to be as simple a criterion for integrality as one could expect. This criterion allows us to pick arbitrary b and d, assuming $b > d$, and then it determines which values of a and c give integral solutions. Again it is particularly striking that a and c only enter via the sum $a + c$. It is also pleasantly surprising that this variant has an easier condition for integrality than the original, though Eq. (11.8) is less symmetric than Eq. (11.1).

In the given problem, $a = c$, $b = 6$, $d = 3$ and Eq. (11.10) simply becomes 3 divides $2a$, or 3 divides a, with the simplest positive answer being $a = 3, x = 45, y = 11$. When $a = c$, the solutions are all multiples of the simplest case. Inspection of small values of b and d shows that the given problem is the smallest "interesting" situation in some sense.

---

[b]This section appeared in *Crux Mathematicorum* 28 (4) May 2002, 236–238.

Bibliography

[1] DIOPHANTOS Book I, prob. 15, pp. 134–135. [See probs. 18 & 19, pp. 135–136, for extended versions.]

[2] T. Došlić, "Fibonacci in Hogwarts," *Mathematical Gazette* 87, 510 (November 2003) 432–436.

[3] Euclid, c. 325 BCE. in *Euclidis Opera Omnia*, vol VIII. J. L. Heiberg & H. Menge, Teubner, 1916, 286–287.

[4] FIBONACCI. Boncompagni, pp. 190–203, 325–346; Sigler, pp. 289–305, 455–480.

[5] D. Singmaster, "Some diophantine recreations." in *The Mathemagician and Pied Puzzler*. ed. by Elwyn R. Berlekamp & Tom Rodgers, A. K. Peters, 1999, 219–235.

Chapter 12

How to Count Your Chickens

> "One rooster is worth five copper cash; one hen is worth three copper cash; three young chicks are worth one copper cash. A man buys 100 fowls with 100 cash; how many roosters, hens and chicks does he buy?"

When Zhang Qiujian concluded his *Suan Jing* [Mathematical Manual] [2] with this problem in c. 475 CE, he little realized that he had created a problem which was soon to be known throughout the world and which would remain a standard problem to the present day and for the foreseeable future. Here we will trace the history of this problem through time and space and some of its many variations.[a]

To start with, you may recognize that this is very like certain "farmyard" problems. For example, a farmer has some cows and chickens in his farmyard and announces that there are 100 heads and 250 feet. How many of each does he have?

If we let x, y be the numbers of cows and chickens, then the statement of the problem gives us the following.

$$x + y = 100,$$
$$4x + 2y = 250.$$

We have two equations with two unknowns. Such problems were already solved by the ancient Babylonians about 1800 BCE and were very well known to the Chinese several centuries before Zhang. With

[a]Written in 1992 for a UK magazine for school students and accepted, but the magazine ceased before this appeared.

modern notation, it is easily done. For example, we can multiply the first equation by 2 to get $2x + 2y = 200$ and then subtract it from the second equation to leave $2x = 50$. Basically, for each equation we can eliminate one unknown from the remaining equations. When we have as many equations as unknowns, we finally reach one equation in one unknown which we can readily solve. In our example, this is the last equation: $2x = 50$, which obviously says that x must be 25. Substituting this into the previous equations, we can successively determine the previous unknowns. So in our example, we substitute $x = 25$ into $x + y = 100$ and find that $y = 75$. Thus our problem has the unique solution $x = 25, y = 75$.

Now consider the 100 Fowls problem. Let x be the number of roosters, y be the number of hens and z be the number of chicks. Then the statement of the problem gives us:

$$x + y + z = 100,$$
$$5x + 3y + z/3 = 100.$$

We again have two equations, but we now have three unknowns. If we try to solve as before, it is easiest to first clear the second equation of fractions to get: $15x + 9y + z = 300$. We now subtract the first equation from this and we have $14x + 8y = 200$, which is one equation in two unknowns. Since we have no more equations, we cannot proceed further with the elimination process and we must consider just what this last equation tells us. Basically it tells us that for each possible choice of x, there is a corresponding value of y. Indeed, we can express y in terms of x by the formula: $y = (200 - 14x)/8$. And we can substitute this into the first equation to express z also in terms of x. Consequently, there are actually infinitely many solutions for this problem.

However, Zhang Qiujian and his successors only considered a few solutions for two reasons. Firstly, they assumed that the numbers x, y, z must be integers. After all, it is a bit messy to get $\frac{2}{3}$ of a chicken! Secondly, they assumed that these numbers must be positive — though later authors are willing to also permit values of zero, so the numbers are non-negative. When both integrality and positivity (or non-negativity) are assumed, then there are only a finite number of solutions of the problem. (Neither assumption is sufficient by itself to restrict the number of solutions to a finite number.)

Most readers may have already made these assumptions and have found these solutions, but let us see how to proceed systematically. We have the equation $14x + 8y = 200$. We can obviously divide out a common factor of two to get $7x + 4y = 100$, and it is not too hard to find the positive or non-negative integral solutions of this. But observe that the coefficient of y and the constant term are both divisible by 4, so that $7x$ and hence x must be divisible by 4. If we let $x = 4x'$, then the equation simplifies considerably to: $7x' + y = 25$. Then x' can take on just 4 values: 0, 1, 2, 3 and the resulting solutions are shown in the table below.

x'	x	y	z
0	0	25	75
1	4	18	78
2	8	11	81
3	12	4	84

There are 4 non-negative solutions and 3 positive solutions. This is denoted by saying there are $(4, 3)$ solutions.

How did Zhang solve the problem? We don't really know. The text gives the last three answers, but only makes the rather cryptic statement: "Roosters increase four each time, hens decrease by seven each time, chicks increase by three each time." This pattern might have been discovered by trial and error, though it may be an observation based on the solutions. But even assuming this pattern was found first, one needs one solution to apply it to and the text gives no indication of how Zhang found his first solution, though he may have just tried letting $x = 1, 2, \ldots$. Notice that Zhang does not give our first solution. Although the Chinese of the time had a good concept of zero, this was based on an empty space on their counting boards and they did not have a symbol for zero until several centuries later. Even when zero had a symbol, it was not a fully accepted value in such problems for over a thousand years. It is just barely possible that Zhang found the first of the above solutions — it is by far the easiest solution and could be found almost by inspection — and then applied his pattern to find the other solutions, but because zero was not an acceptable value, he then omitted any mention of it.

Zhang Qiujian may have invented this problem. The facts are that we do not know of any earlier appearance of the problem, but certain similar problems were known earlier, both in China (a problem with five equations in six unknowns was done about 500 years earlier) and in Greece (Diophantos of Alexandria gives general solutions of some problems about 250 CE). In such a situation, it is impossible to be definite. Even if Zhang had specifically asserted that he had invented the problem, one could have doubts — there are many examples of people claiming inventions that they did not make. However, the 6th and 7th century Chinese successors of Zhang showed considerable incompetence in trying to explain Zhang's solution and in trying to create new versions of the problem. Their methods tended to be just guesswork and they did not always find all the positive solutions. This lends considerable support to the belief that Zhang's work was original.

At this point, we come across a spanner in the historical works. In 1881, a collection of birch-bark pages was uncovered near the village of Bakhshālī in the Peshawar district of India, near the present border with Pakistan. Upon examination, it turned out to be a mathematical work, but severely damaged. Since then, great controversy has raged over this Bakhshālī Manuscript [3], with datings ranging from second century BCE to 12th century CE making it the most uncertainly dated of all ancient mathematics texts. Historians seem to be converging toward a date of either 4th century or 10th century — Indian historians tend toward 4th century; David Pingree, the leading Western historian of Sanskrit mathematics, opines 10th century; the most substantial study of the MS is by Pingree's student Takao Hayashi and he opines 7th century. The sophistication of some of the problems inclines me to this date. Three of the problems are of the 100 Fowls type, but one is essentially lost and the second is only partly readable. The third is a clear example of our problem, but in a different setting.

> A man earns 3 mandas in a day, a woman $1\frac{1}{2}$ mandas in a day, and a sudha $\frac{1}{2}$ manda in a day. If 20 of them earn 20 mandas in a day, how many of each category are there?

(The number of solutions will be given at the end of this chapter.)

We have seen that the problem appears to have originated in China in the 5th century and to have continued there in the 6th and 7th centuries, with an occurrence in India perhaps about that time. Then the problem appears almost simultaneously in India, Egypt and Northern Europe in the 9th century. In each case, the problem is given in several versions, with various extensions, so that the problem must have been well known at the time it was written down. When ideas or problems passed from the Orient to Europe, there was usually a gap of a few hundred years between the stages — the 100 Fowls problem is remarkable for the rapidity of its diffusion. Let us now look at these occurrences.

Surprisingly, the Northern European versions are the earliest recorded examples among these. Alcuin of York (c. 732–804), discussed in Chapter 3, had amongst his 53 problems, no less than 8 versions of the 100 Fowls problem. Three of these involve buying 100 animals for 100 pence or shillings. The first two of these involve buying: boars, sows and piglets; horses, cows, sheep. The third is as follows.

> 39. A man in the East wanted to buy 100 assorted animals for 100 shillings. He ordered his servant to pay 5 shillings for a camel, one shilling for an ass and one shilling for 20 sheep. How many camels, asses and sheep did he buy?

The other five problems involve distributing grain to men, women and children or bread to priests, deacons and readers, of which the following is an example.

> 34. A gentleman has a household of 100 persons and proposes to give them 100 measures of grain, so that each man should receive three measures, each woman two measures, and each child half a measure. How many men, women and children are there?

If you are solving these, then you will have noticed that Alcuin's 34 has several solutions. However Alcuin only gives one solution to each problem and merely verifies that it works. He shows no sign of knowing that more solutions can exist.

We now shift to India, which was in the middle of a great period of mathematics, represented here by the works of Mahāvīrā (850), Ṡrîdharâ (c. 900) and Bhāskara II (1150). Mahāvīrā gives some general techniques for the problem and three examples. The first involves buying 68 palas of drugs for 60 panas, where the drugs are ginger, long pepper and pepper worth $\frac{3}{5}$, $\frac{4}{11}$ and 8 panas per pala. The other two involve buying four types of fowls: peacocks, pigeons, swans and sârasa birds. The first buys 72 birds for 56 panas with prices $\frac{2}{3}, \frac{3}{4}, \frac{4}{5}, \frac{5}{6}$ panas per bird. The second is below.

> Pigeons are sold at the rate of 5 for 3, sârasa birds at the rate of 7 for 5, swans at the rate of 9 for 7, and peacocks at the rate of 3 for 9. A certain man was told to bring at these rates 100 birds for 100 panas for the amusement of the king's son, and was sent to do do. What does he give for each? [How much does he pay for each type of bird?]

Both Śrîdharâ and Bhāskara II repeat this last problem, only slightly changing the types of birds (though this may just be due to different translators). (See Chapter 3.)

It should be clear that the problems with four types of bird are substantially harder than with three types, even without the complication of formidable denominators. If we let w, x, y, z be the numbers of birds, we have:

$$w + x + y + z = 100;$$
$$\frac{3}{5}w + \frac{5}{7}x + \frac{7}{9}y + 3z = 100.$$

Clearing the fractions from the second gives us:

$$189w + 225x + 245y + 945z = 31500.$$

Subtracting 189 times the first equation from this leaves us with:

$$36x + 56y + 756z = 12600.$$

We can divide out a factor of 4, obtaining:

$$9x + 14y + 189z = 3150.$$

Since 9 divides 9, 189 and 3150, it must also divide $14y$ and hence y, so we set $y = 9y'$ and then we can cancel a factor of 9 to get:

$$x + 14y' + 21z = 350.$$

Now we similarly see that x must be divisible by 7, so we set $x = 7x'$ and obtain:

$$x' + 2y' + 3z = 50.$$

Despite the substantial simplifications, this is just *one* equation in *three* unknowns and finding all the solutions is a tedious process, which none of the Indian authors bothered with — they all give just one solution with no indication that there might be others, though a later commentator on Śrîdharâ gives (4, 4) solutions of this problem as well as (5, 5) solutions of Śrîdharâ's other problem: to buy 100 fruits for 80 rupas when pomegranates cost 2, mangoes cost $\frac{3}{5}$ and wood-apples cost $\frac{1}{2}$.

Recall that Zhang's problem led us to one equation in two unknowns: $7x' + y = 25$. If we plot this on a graph, it is the line which cuts the x'-axis at $\frac{25}{7}$ and the y-axis at 25. The points of interest are the integral points (i.e. both coordinates are integers) which are on this line and which lie between the two axes. (Whether the points on the axes are included or not depends on whether one permits zero values or not.) Because these points are regularly spaced, it is not too difficult to count them — the complications arise at the ends. But our equation $x' + 2y' + 3z = 50$ represents the plane which cuts the x'-axis at 50, the y'-axis at 25 and the z-axis at $\frac{50}{3}$. The points of interest are the integral points which lie within this triangle. From this we appreciate why Mahāvīrā's problem is so much more difficult to resolve. Indeed the only way to proceed is by trying each value of one unknown and solving the resulting equation in two unknowns. In our example, we can try $z = 0, 1, \ldots, 16$. Having done this a few times, one soon wishes for a computer program to do it. All the examples have been checked by computer.

Our attention now shifts to the Arabic world where Abu Kamil Shuja ibn Aslam ibn Muhammad ibn Shuja "al-Hasib al-Misri", generally known as Abu Kamil "the reckoner from Egypt", wrote a short work entitled *The Book of Rare Things in the Art of Calculation* c. 900. (See Chapter 4 problem 39.) He begins thus.

> In the name of God, the merciful, the compassionate. So speaks Shuja ibn Aslam, known by the name of Abu Kamil: I know a particular kind of reckoning, which circulates among the distinguished and the ordinary, among learned and unlearned, which delights them and which they find new and beautiful; if one asks others [about their solution] then he will be answered with an inexact, suppositious answer, in which they recognize neither a principle nor a rule. Many distinguished and many ordinary people used to ask me about problems in calculation and I answered them with the unique answer for each problem, if

> there were no other answers; but often there were, for one problem, two, three, four and more answers, so often one answer was impossible. Yes, I even succeeded with a problem which I solved and for which I found very many solutions; I examined the thing thoroughly and came to 2676 correct solutions![b] My astonishment was great, and I had the experience that when I described this discovery, I was regarded with astonishment or considered crazy, while those, who did not know me, formed a false suspicion against me. So I resolved to write a book about this kind of reckoning in order to simplify the process itself and to bring better understanding. This I have now begun, and I will explain the solutions for those problems which have several solutions, and for those which have only one, and for those which have no solutions, with the help of a certain method; finally I will treat the problem, which I have mentioned, that has 2676 solutions. So the suspicion and the presumptions will vanish away, and my statements will be confirmed and the truth will appear. If I should add yet more thoughts, about these and similar problems, which may occur to me regarding the great number of solutions, then the information will be distributed.

Abu Kamil then proceeds to give six examples of the 100 Fowls problem, each involving buying 100 fowls for 100 drachmas. The first has ducks, hens and sparrows costing 5, 1 and $\frac{1}{20}$ drachmas per bird. If you have been paying attention, you will notice that this is numerically identical to Alcuin's problem 39 which is quoted above. This leads to speculation discussed later. The second has ducks, doves, hens, costing 2, $\frac{1}{3}$, $\frac{1}{2}$. The third has ducks, sparrows, doves and hens, costing 4, $\frac{1}{10}$, $\frac{1}{2}$, 1. Abu Kamil makes some errors and only finds (96, 96) of the solutions — a later commentator found the missing ones. His fourth problem has ducks, doves, larks and hens, costing 2, $\frac{1}{2}$, $\frac{1}{3}$, 1. The fifth problem is notable — it has ducks, hens and sparrows costing 3, $\frac{1}{3}$, $\frac{1}{20}$. He states that this has no solution, which is correct for his context of positive solutions, though there is one non-negative solution: 25, 75, 0. The sixth is his masterpiece. He has five types of bird: ducks, doves, ring-doves, larks and sparrows, costing 2, $\frac{1}{2}$, $\frac{1}{3}$, $\frac{1}{4}$, 1. He says there are 2676 solutions once and 2696 solutions twice, but the latter are considered to be scribal errors. But

[b]Here the text reads 2696, but this is a scribal error.

even 2676 is wrong, though this doesn't seem to have been pointed out until the 1960s.

From Abu Kamil's introduction, the complexity of his examples and the fact that he finds all solutions, it seems clear that the problem must have been well known for some time in the Arabic world. The prevalence of birds and the increasing sophistication of the problems make it seem likely that the problem diffused from China, through India, to the Arabs, carried by travelers now obscured by time. The numerical identity of Alcuin's problem 39 and its context also make it seem likely that Alcuin had acquired it from some Near-Eastern source, perhaps through Byzantium, but again we know nothing of such contact. It has been suggested that these problems show that Alcuin's *Propositiones* must be derived from Abu Kamil and hence must be much later than Alcuin's death in 804, but the earliest manuscript seems to really be 9th century, this is about contemporary with Abu Kamil. It seems more likely that both were drawing on earlier examples. Perhaps some new document will appear in a cave in China or a monastery in Greece or a temple in Afghanistan and clarify the situation.

After the spectacular blooming of the 100 Fowls problem in the 9th century, it has remained a standard problem, appearing regularly in all parts of the world. A few of its later appearances will now be discussed.

Al-Karkhi (= al-Karagi), in c. 1010, [4] gives an example of a very similar problem — it has much older antecedents, but are not discussed here. One has a certain type of goods (the translator leaves it in Arabic) of different qualities, worth 5, 7, 9 dirhams (per unit). How can one mix these to produce one unit worth 8? This leads us to the following:

$$x + y + z = 1,$$
$$5x + 7y + 9z = 8.$$

However, we immediately see that there is no reason for x, y, z to be integers — indeed there is no way they can be integers. There are infinitely many solutions. Al-Karkhi is happy to find just one, which he does by assuming $x = y$.

Fibonacci, in 1202, devotes an entire chapter, 23 large pages in the printed edition, to such problems, under the heading of "consolidation of money," referring to the process of combining metals of different values to produce a result of a desired value. Sometimes the total to be produced is given, other times it is not. If the total were not given, al-Karkhi's problem becomes:

$$5x + 7y + 9z = 8(x + y + z).$$

Although integral solutions are not needed, Fibonacci usually gives one or a few integral solutions. He gives examples with 7 and 240 values of metal, but without integral solutions. He gives a notable example of the 100 Fowls type — a man buys 7 pounds of meat for 7 pence, pork costs 3, beef costs 2, and hyrax (an animal somewhat like a rabbit) costs $\frac{1}{2}$. This has no non-negative integral solutions. He gives 1, $\frac{2}{3}$, $5\frac{1}{3}$. He gives an example with four types of bird, but only gives 2 of the 19 positive solutions. In a letter, Fibonacci gives a problem of buying 15 birds for 15 dinars, with sparrows worth $\frac{1}{3}$, turtledoves worth $\frac{1}{2}$ and pigeons worth 2. This has (0, 2) solutions. He says this cannot be solved without fractions, but notes that 0, 10, 5 solves the problem, and then that, "if we want to break birds," one can have $\frac{9}{2}$, 5, $\frac{11}{2}$.

From this time on, both the 100 Fowls and the mixing problems are common, but it is rare for more than one solution to be given until the 18th century. For example, a Lucca manuscript of c. 1390 has a problem of mixing five kinds of metals which has (3027289, 2966486) integral solutions, but only one is given. The most interesting exception is Nicolas Chuquet, writing in 1484. A merchant bought 30 ells of cloth for 30 ecus. There were three types of cloth, costing 4, 2 and $\frac{1}{2}$ ecus per ell. He gives the unique positive solution: 3, 3, 24, but he then gives 4, $\frac{2}{3}$, $25\frac{1}{3}$ and says: "And in this manner one can make as many answers as one wishes." Elsewhere, he says "Also one should know that all such calculations have several answers, as many as you want, as appears [later in] this book." He also gives a two variable problem with a negative solution and tries to interpret this.

From about 1400, the context of the 100 Fowls problem was often changed to men, women and children paying a bill at a tavern and texts give a rule for it, variously "regula coecis" or "regula Virginum", while the mixing problems are called "Gold- und Silberrechnung," "zech rechnen" or "regula alligationis" — "alligation" was a common topic in English textbooks until about 1900. Tartaglia, writing in 1556, gives a solution with a zero value in a problem with no positive solutions, but he then rejects it and gives a fractional solution as in a previous problem. In that previous problem he said that a fractional solution isn't really acceptable because the problem specifies live birds. However, the problem where he gives a zero value involves buying grains and there would then be no reason to reject fractional answers.

In the 18th century, a new variation of the problem appears, involving money. The earliest seen are in Thomas Simpson's *A Treatise of Algebra* of 1745. A man wishes to pay a debt of £100, using guineas and pistoles. Clearly one must know that a pound (£) was 20 shillings (s), a guinea was 21s and a pistole was 17s. Then the problem becomes:

$$21x + 17y = 2000,$$

which is one equation in two unknowns and has several solutions. So here we get the same kind of problem as arose after elimination in the 100 Fowls problems, but there we began with two equations in three unknowns. In Simpson's next problem a man wants to pay £100 using guineas (worth 21s) and moidores (worth 27s). In the 1800 edition of this book, the problem is extended as far as actually describing all the solutions of

$$12x + 15y + 20z = 100,001.$$

The English edition of Euler's famous algebra text of 1770 gives a number of general techniques and some examples using money, such as the following:

> I owe my friend a shilling, and have nothing about me but guineas, and he has nothing but louis d'ors, valued at 17s. each; how must I acquit myself of the debt?

In the 19th century and 20th century, more versions with money appear. For example: change a pound (=20s) into 18 coins comprising half crowns ($= 2\frac{1}{2}$s), shillings (=1s) and six-pences ($= \frac{1}{2}$s). Also: change a shilling (=12d) into 12 coins, without using pennies (d).

Hence one could use six-pences (= 6d), three-pences (= 3d), half-pennies (= $\frac{1}{2}$d) and farthings (= $\frac{1}{4}$d). How would this change if one allowed pennies?

Perhaps the most peculiar example occurs in [1]. We want to use 26d in florins (= 24d), shillings (= 12d), six-pences (= 6d), pennies (= 1d) and half-pennies (= $\frac{1}{2}$d) to measure a distance of $5\frac{5}{8}$ inches. The widths of the coins, in 16ths of an inch are: 18, 15, 12, 19, 16, respectively.

To show that such problems are indeed alive and well, here is one of my *Weekend Telegraph* "Brain Twisters."

> Travelling through California, one expects to run across odd things, but I was more than a bit intrigued to see a sign post: THE CCCC RANCH. So I drove off along the indicated side road and about five miles later I found the ranch, which was really more of a small farm. A young man came out and asked our business. When I told him it was just curiosity about the name, he guffawed, "Lotsa folks wonder about that. Well, we raise four kinds of animals that begin with C." "I see," I replied, "if you'll excuse the pun, a cow and a chicken already. What are the other kinds?" "Children and copperheads—we found so many of the damn snakes on the land that we've gone into raising them for their venom."
>
> My daughter Jessica piped up, "How many snakes have you got?" "Do you like puzzles?" "Oh, yes!" "Well then, at the moment we've got 17 heads all together." "Do you have any two-headed animals?" "Nope, but that's a smart question. The animals have 11 tails." "Do you count people as having tails?" "Nope, these are all genuine tails and there ain't any double tailed beasts either!" "How many legs have they got and are they all normal?" "They've got 50 legs and they're all normal." "Let me see," said Jessica, "there are four unknown numbers, so I need some more information." "Sorry," said the lad, "that's all I'm a'gonna tell you, except that we've got some of each." Can Jessica find the number of animals of each kind?

The 100 Fowls problem is such a clearly identifiable problem that it serves as a kind of historical tracer, identifying the diffusion of mathematics in space and time. When combined with other distinctive problems such as the Chinese Remainder Theorem, we get a very strong impression of the amount of mathematics which came from China, via India and the Arabs to Europe, a cultural transmission

Jtem/einer hat 100. fl. dafür wil er 100. haupt Vihes kauffen / nemlich / Ochsen/ Schwein/ Kälber/ vnd Geissen/ kost ein Ochs 4 fl. ein Schwein anderthalben fl. ein Kalb einen halben fl. vnd ein Geiß ein ort von einem fl. wie viel sol er jeglicher haben für die 100. fl? Machs nach den vorigen/mach eines jeglichen kosten zu örtern/deßgleichen die 100. fl. vnd setz als dann also:

	16	15	
	6	5	
100			400
	2	1	
	1		

Multi.

Figure 12.1. This is the only early illustration found for this problem, from: Adam Riese, *Rechnung auff der Linien unnd Federn* .... This first appeared in Erfurt, 1522, and was reprinted in 1544 and 1574, but the illustration only occurs in the latter: Christian Egenolff's Erben, Frankfurt, 1574 (the text is dated 1525). Facsimile by Th. Schäfer, Hannover, 1987, p. 71r.

which has not been recognized until quite recently. So the 100 Fowls problem has a great historical significance and it is good fun.

12.1 Answers

Only the numbers of solutions are given, first non-negative and then positive.

Bakhshali Manuscript: (3,1).
Alcuin 39: (2,1).
Alcuin 34: (7,6).
Mahāvīrā, first: (1,1).
Mahāvīrā, second: (217,169).
Chuquet: (2,1).
Simpson, first: (6,6).
Simpson, second: (0,0).
Simpson, third: (1388611,1388611).
Euler: (1,1).

Mahāvīrā, third: (26,16).
Ṡrîdharâ's other: (6,5).
Abu Kamil 1: (2,1).
Abu Kamil 2: (7,6).
Abu Kamil 3: (122,98).
Abu Kamil 4: (364,304).
Abu Kamil 5: (1,0).
Abu Kamil 6: (3727,2678).
Changing money, first: (4,4).
Changing money, second: (2,1).
Last modified to permit pennies: (13,0).
Adams: (1,0).

Singmaster: Yes, the numbers are 9, 1, 6, 1 of cows, chickens, children and copperheads. There is also a solution with values 8, 3, 6, 0.

Bibliography

[1] M. Adam, *The Morley Adams Puzzle Book.* Faber and Faber, 1939.
[2] Z. Qiujian. *Zhang Qiujian Suan Jing* [Zhang Qiujian's Mathematical Manual]. 468. Chap. 3, last problem.
[3] BAKHSHĀLĪ.
[4] AL-KARAGI.

Chapter 13

The Monkey and the Coconuts

In the 9 October 1926 issue of *The Saturday Evening Post*, the American writer Ben Ames Williams had a short story called "Coconuts" [36]. The story centers around the **monkey and the coconuts problem**, as described here.[a]

> Five men and a monkey were shipwrecked on a desert island, and they spent the first day gathering coconuts for food. Piled them all up together and then went to sleep for the night. But when they were all asleep one man woke up, and he thought there might be a row about dividing the coconuts in the morning, so he decided to take his share. So he divided the coconuts into five piles. He had one coconut left over, and he gave that to the monkey, and he hid his pile and put the rest all back together. So by and by the next man woke up and did the same thing. And he had one left over, and he gave it to the monkey. And all five of the men did the same thing, one after the other; each one taking a fifth of the coconuts in the pile when he woke up, and each one having one left over for the monkey. And in the morning they divided what coconuts were left, and they came out in five equal shares. Of course each one must have known there were coconuts missing; but each one was guilty as the others, so they didn't say anything. How many coconuts were there in the beginning?

[a]Appeared, without the illustrations, in "Gaṇita Bhāratī" *Bull. Ind. Soc. Hist. Math.* 19 (1997) 35–51. Based on Technical Reports SBU-CISM-93-02, SBU-CISM-96-08 and SBU CISM-97-06, School of Computing, Information Systems and Mathematics, South Bank University, February 1994, January 1996 and September 1997. Presented at 2nd Gathering for Gardner, Atlanta, 19 January 1996.

In Williams' story, this problem is used as a ruse to distract one of the protagonists long enough that he doesn't submit a bid for a contract. From the time the problem is introduced, the story describes the attempts made to solve the problem and, at the end, it is still unsolved. This produced an avalanche of letters, telegrams and phone calls at the *Post* office. Martin Gardner describes this in one of his early articles (April 1958) [17] and states that about 2000 letters arrived in the first week, leading the editor to telegram Williams as follows:

FOR THE LOVE OF MIKE, HOW MANY COCONUTS?
HELL POPPING AROUND HERE.

Perhaps Williams had no idea of the answer. Maybe someone can look up Williams' life to see if he had any mathematical training. Martin Gardner describes the problem as the "most worked on and least often solved of all the Diophantine brain-teasers." If you have never done this problem, try doing it now. Gardner tried to trace the problem to its origins and got back to 1912 [31]. He found that there were two versions of the problem, depending on whether the final division was exact or not, that is, whether the monkey does or does not get a final coconut.

What is a Diophantine problem? This can mean different things in different contexts. The reference is to the style of problems that Diophantus gave many examples of, see Chapter 2. It is understood that we are looking for integral solution(s) to one or more equations. (Curiously Diophantus did not always constrain himself to integers!) When there are enough constraints that there is exactly one solution (or provably none) the problem is "determinate." Generally when there are more variables than equations there can be two or more solutions and the situation is "indeterminate," and many people reserve the term "Diophantine" for the indeterminate situation.

13.1 Determinate Versions

There are much older versions of the problem. Here is a version from the *Trattato d'Aritmetica* of c. 1370, attributed to Paolo dell'Abbaco [1].

> A man sent one of his children to a garden to fetch seven apples, saying: "You will find three gatekeepers, each of whom will say: 'I want half of all your apples and two more from those which remain after the division.' I want to know how many have to be taken at the beginning, so that seven will remain at the end."

One could phrase this as: (give half and two more) thrice to leave 7. Problems of this type are determinate and are easily solved by working backward from the end.

Here we begin with the 7 and (add two and double) thrice to get the numbers: 7, 9, 18, 20, 40, 42, 84, and one must start with 84 apples.

Apple (or other fruit) stealing (or collecting) versions have been popular since the Middle Ages, in many variations. One can change the fraction, the extra amount and the number of stages. Fibonacci, 1202, [15, p. 278] gives an apple version with seven steps: (give half and one more) seven times to leave 1. A later popular version usually involved selling eggs and the vendor said she had sold half her eggs and half an egg more three times, leaving 36 [26, 27]. An initial reaction is that she must have sold some half-eggs! For determinate problems, one can easily allow the fraction and the extra amount to vary. For example, John Jackson's *Rational Amusement for Winter Evenings* of 1821 gives the following [19].

> A gentlemen gave to the first of three poor persons that he met, half the number of shillings that he had about him, and one shilling more; to the second, half of what remained, and two shillings more; and to the third, half what now remained, and three shillings more; after which he found he had only one shilling left. How many shillings had he at first?

Fibonacci [15] gives an extended version: A man leaving a city has to pay a toll at each of ten gates. At the first, he pays $\frac{2}{3}$ of his money and $\frac{2}{3}$ of a unit more. Then at the ith gate, he pays $\frac{1}{i}$ of his money and $\frac{1}{i}$ of a unit more, for $i = 2, 3, \ldots, 10$, leaving him with 1. (This problem occurs in an earlier untranslated Persian source, Ṭabarī c. 1075, [33].)

There are earlier forms of the problem, usually involving merchants traveling to a number of fairs. An early version occurs in one of the more exotic sources, the *Arithmetical Problems* by Ananias of

Figure 13.1. dell'Abbaco, [1], c. 1370, Ragione (= problem) 47, p. 45. The first problem given in Section 2 above.

Shirak, an Armenian manuscript of c. 640 containing 24 problems [3]. His problem 19 is as follows.

> A man went into three churches and prayed to God the first time: "Give me as much as I have, and I will give you 25." He did the same the second time, giving 25, and the same the third time. And then he had nothing left. Now how much did he have to start?

The late Heinrich Hermelink, commenting on the history of this problem [18], noted that here the doubling is due to prayers in churches, but then the Arabic world converted it to prayers in mosques, then the West converted back to churches, but in the Renaissance, the doubling was due to gambling. Most Renaissance versions are actually due to profit at trading and indeed this is the basis of the earliest known examples in the *Chiu Chang Suan Ching* [10], the first major mathematical text in China, dating from somewhere about 100 BCE to 100 CE. One of the problems [10] has a merchant making 30% profit at each of five fairs, and after the fairs, he sent home 14000,

Figure 13.2. dell'Abbaco, [1], c. 1370, Ragione 71, p. 66. This has a man passing through three doors and (losing 1/3 and 6 more) thrice to leave him with 24 — no objects are specified, but he seems to be holding a money bag.

13000, 12000, 11000, 10000, respectively, leaving him with nothing after the fifth fair. How much had he to start? Although determinate, anyone who does this question will be impressed by the arithmetical ability of the ancient Chinese.

Another problem in the *Chiu Chang Suan Ching* [10] shows how the problem probably developed from simpler problems. A man carrying rice has to pay $\frac{1}{3}$, $\frac{1}{5}$, $\frac{1}{7}$ of his load at three customs points, and he is left with 5. At this point, one is nearly back to some of the oldest problems of all — the "aha" or "heap" problems which already occur in the *Rhind Papyrus* of c. 1700 BCE [30]. For example, no. 24 and no. 28 are

> "A quantity and its $\frac{1}{7}$ added together become 19. What is the quantity?"
>
> "A quantity and its $\frac{2}{3}$ are added together and from the sum $\frac{1}{3}$ of the sum is subtracted, and 10 remains. What is the quantity?"

Some versions of the problem give not the remainder, but the amount paid out. For example, another problem in the *Chiu Chang Suan Ching* [10] has a man paying $\frac{1}{2}$, $\frac{1}{3}$, $\frac{1}{4}$, $\frac{1}{5}$, $\frac{1}{6}$ of his amount, making a total of 1 paid out. Sometimes the end amount is given in terms of the original amount. For example, in the *Bîjagaṇita* of Bhāskara, 1150 [7], we find the following.

> A trader, paying ten upon entrance into a town, doubled his remaining capital, consumed ten [during his stay] and paid ten on his departure. Thus, in three towns [visited by him] his original capital was tripled. Say what was the amount?

Fibonacci gives a similar problem [15] and he also gives a number of problems where the starting amount is given and the amount spent has to be determined, but these usually immediately follow the identical problem with the amount spent given and the starting amount determined. However, he gives one extraordinary problem to find the number of stages: start with 13 (double and then spend 14), n times to leave 0. He gets $n = 3\frac{3}{4}$ times by a linear interpolation between 3 and 4. It should be $n = \log_2 14 \approx 3.80735$ times.[b]

Chuquet, 1484 [11] (prob. 95) [16] (p. 219), gives an even more difficult problem.

> A merchant has made as many journeys as he had pieces of gold, in such a way that on his first journey he made from 3, 4, and 1 more. On the second journey, he made from 3, 4, and two more. On the third journey, he did as above, and three more. And thus continuing, always making from 3, 4, and progressing by 1, so that at the end of his journeys, 15 pieces of gold are found. To determine how many pieces he had at the beginning.

That is, he starts with X and gains $\frac{1}{3}$ of his capital and i units more on his ith journey, producing 15 after X journeys. Letting $r = \frac{4}{3}$ and X_n be the amount after n journeys, we get

$$X_n = r^n X_0 + \frac{r^{n+1} - (n+1)r + n}{(r-1)^2},$$

so if $n = X_0 = X$, we get that X is the solution of $(X + \frac{64}{3})r^X - 3X = 27$. Chuquet gives an elaborate interpolation leading to $X \approx 3.03949414$, but we get $X \approx 3.045827298$.

[b]Problems like this which lead to interpolation into a geometric progression are rare before the development of compound interest during the Renaissance. The only other examples seem to be in the *Chiu Chang Suan Ching* [10].

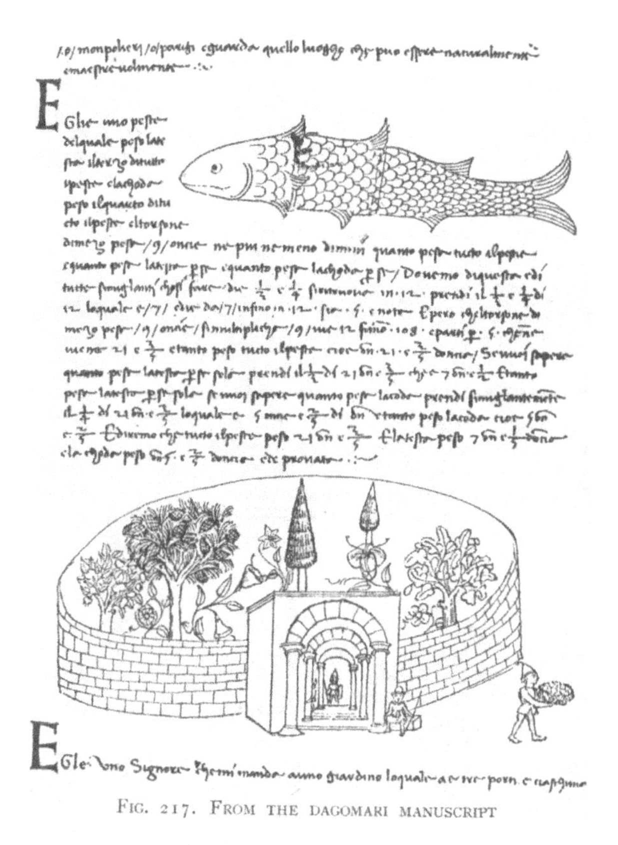

Figure 13.3. From the Dagomari Manuscript, an MS attributed to dell'Abbaco [2]. The reproduction shows only one line of text for this problem and it says: "There is a man who sends to a garden where there are apples ...," but there are no numerical data shown.

Euler's *Algebra* (or at least the 1770 English edition of it [14]) gives a different formulation. A man (spends 100 and then gains a profit of $\frac{1}{3}$ on his remainder) thrice to double his original money. That is, the subtraction is done before the division which is rare in determinate problems, but common in indeterminate problems. L. Mittenzwey's *Mathematische Kurzweil* of 1879 has yet a different form [24], which may be the background to the coconuts problem. Three men successively take $\frac{1}{3}$ from a pile of potatoes, leaving 24. How should these 24 be distributed to make everyone equal?

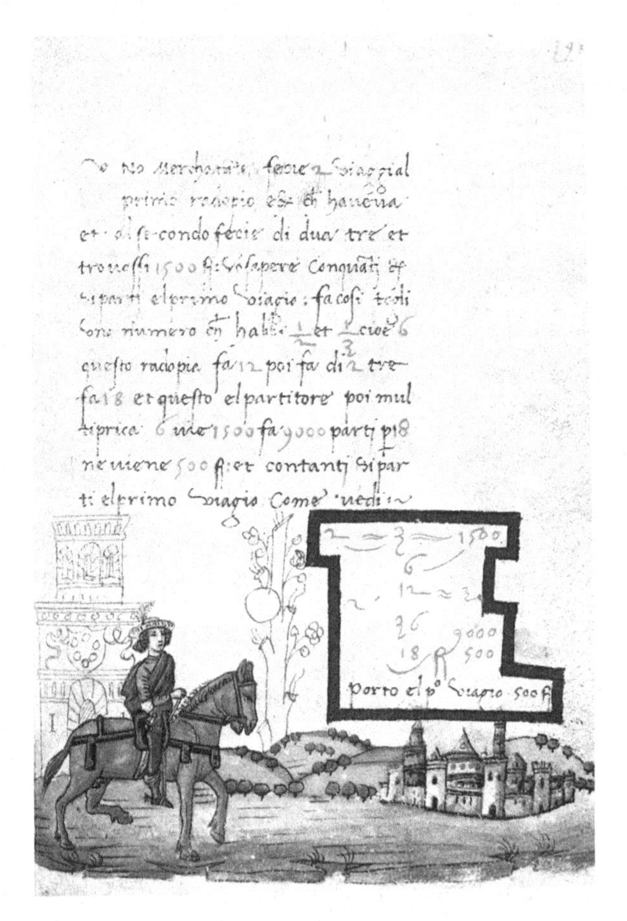

Figure 13.4. From the *Calandri Aritmetica* [9, f. 98b], a beautifully illuminated MS prepared for the Medici family c. 1500. A merchant travels to two fairs. He doubles his money at the first and 3/2's it at the second, resulting in 1500. How much did he have to start?

13.2 Indeterminate Versions

Having discovered all these determinate problems, one would assume that the modern indeterminate versions must have developed from them in some way and there would find some signs of the development in the 19th century, leading up to the 1912 work [31] cited by Gardner. The 1912 item was the solution of a problem in *School Science and Mathematics*, but it only did a coconuts problem for three men with the final division leaving one for the monkey. None of the three solvers considered the general problem. But one of the solvers, Norman H. Anning, pointed out that the solution is $-2 \pmod{3^4}$, so -2 is a solution! The first general solution seen is in 1928 [34], for the case where the final division is exact.

A. Cyril Pearson has some nice versions. One version [28] (no. 29) has three men with a monkey who gather mangoes, but on the next day when they divide the final pile in thirds, there is an extra mango left over for the monkey. Another version [28] (no. 94) has a bag of nuts being divided between four boys who have a parrot, again with an extra for the parrot at the final division. For several years, these were the earliest modern versions of the coconuts problem known.

However, in late 1996, Jerry Slocum provided a copy of Clark's booklet of puzzles [12] of 1904, which says its first edition was 1897. This has several determinate versions and an indeterminate version, Number 99: "Three Boys and Basket of Apples," where an extra apple is thrown away at each stage and at the end. So this is now the earliest known modern example and the earliest known example with one left over at the end. But Clark's booklet is really a collection of old chestnuts and seems unlikely to contain much, if any, new material, so earlier versions probably exist, but they may be hard to locate.[c]

From about 1930 onward, the indeterminate problem appears regularly, with varying persons (shipwrecked sailors, boys, Italian organ grinders, explorers), varying companions (monkey, parrot, cat, man Friday) and varying objects (apples, coconuts, mangoes, nuts, peanuts, doubloons, biscuits, pears), and then with additional variations.

A few earlier versions had turned up, but these do not seem to connect to the modern versions. Dudeney has a version in his *Weekly Dispatch* column in 1903 [13] where the initial amount of eggs is a multiple of 25 and one (sells half the eggs and half an egg more) until they are all gone. Montucla's edition of Ozanam, 1778 [27], and Lucas, 1895 [22], had worked out that if one starts with $2^n - 1$ eggs, then one can (sell half the eggs and half an egg more) n times to

[c]In 2000, Jerry Slocum sent several other versions of Clark, showing that there were at least ten editions of it, with considerable differences. The problem of interest is not in the 1897 edition and the earliest version known to contain it is the 1904 edition.

leave 0. The 1725 edition of Ozanam [26] actually made the more general statement that $8a - 1$ eggs will leave $a - 1$ and that one can replace 8 by 2^n if one does the process n times.

Surprisingly, two such "coconuts" problems are found in the *Gaṇita-sāra-sangraha* of Mahāvīrā [23], which dates from 850! This book has a very nice collection of puzzles. It one of the most readable of ancient texts. It was translated into English and published in Madras in 1912, with a reprint in 2000. Reading the problem statements, the author seems to have been writing about a century ago. However the solutions are pretty hard to follow because of the lack of algebra. Fortunately, the translator has generally explained the solutions in modern algebraic notation, but they are still often rather roundabout and unclear to a modern reader.

Let me quote the problems in full (material in parentheses is amplification by the translator).

> On a certain man bringing mango fruits (home, his) elder son took one fruit first and then half of what remained. (On the elder son going away after doing this), the younger (son) did similarly (with what was left there. He further took half of what was thereafter left); and the other (son) took the other half. (Find the number of fruits brought by the father.) [23]

That is, we have the problem: (take one and then half of the remainder) twice to leave a multiple of two. The division process is identical to Euler and the usual indeterminate problems — take away some first, then divide into equal parts, but the amount taken is taken by the person rather than given away to the monkey. The ending is identical to Williams' 1926 version, i.e. the final division is exact.

> A certain person went (with flowers) into a Jain temple which was (in height) three times the height of a man. At first he offered one (of those flowers) in worship at the foot of the Jain and then (offered in worship) one-third of the remaining number (of flowers) to the first height-measurement (of the Jain). Out of the remaining two-thirds (of the number of flowers, he conducted worship) in the same manner in relation to the second height-measure; and (then he did) the same thing in relation to the third height-measure also. The two-thirds which remained at last were also made into 3 equal parts (by him); and having worshipped the 24 *tīrthaṅkaras* (with these parts at the rate of eight *tīrthaṅkaras* for each part) he went away with no (flower) on hand. (Find out the number of flowers taken by him.) [23]

The somewhat flowery language in the translation makes it a bit hard to see, but this is: (give one and then a third of what is left) thrice to leave a multiple of three.

Mahāvīrā gives a number of other problems, but these are all determinate. The presence of mangoes in Mahāvīrā and in Pearson leads to interesting historical conjectures. Was there any connection between Mahāvīrā and Pearson? The translation of Mahāvīrā did not appear until 1912 and Pearson's book was published in 1907, but it is based on his puzzles in the *Evening Standard* over the previous two years. The frontispiece photo of Pearson [28] shows him as a clergyman of about 40–50 years old, and another of his books describes him as MA of Balliol College, Oxford, and rector of Drayton Parslow, Buckinghamshire, with a stamp underneath giving Springfield Rectory, Chelmsford [Essex], so he might have been a missionary in India in his younger days or had Indian contacts. Perhaps Mahāvīrā's translator had published some of the problems while he was working on the translation. Perhaps the manuscript of the translation circulated before publication — indeed D. E. Smith describes the work in 1908/1909 [32]. Perhaps Mahāvīrā's work, or at least this problem, was well known in India — the problem might even be part of Indian folk-mathematics. In [20], we find a reference to [23] as 1908 with the note: "This is really an advance copy of a work not yet actually published, kindly supplied to me by the author [i.e. the translator]." It would be interesting to find out something about Pearson — from the DNB and the evidence in one of his books, he was Arthur Cyril Pearson, rector of Springfield, Essex, and the father of Cyril Arthur Pearson (1866–1921), the newspaper and magazine publisher, later knighted and baroneted. In particular, did Pearson have any Indian connections? However, the lack of these is not very definitive, as he might have had a friend with Indian connections. Pearson may have had a letter from some (Indian?) correspondent suggesting the problem. Unless more evidence turns up, we cannot know just how this 9th century Indian problem resurfaced in England in 1907, though the evidence in [20] has considerably filled in the gap.

However, the recent discovery of the Clark example rather puts a spanner into the above speculations and unexpectedly opens up a new region for investigation, though it is still very possible that the translator of Mahāvīrā had published or circulated parts of the translation. We know D. E. Smith had seen the whole translation by

c. 1908. Perhaps he had some connection with Clark or perhaps he published the problem in some American journal?

W. W. Rouse Ball [6] has a version with three Arab jugglers, traveling to Mecca with their performing monkey, who buy a basket of dates. The problem is Form 1 and Ending 1, but he initially states that the original basket has one more than a multiple of three in it, though this is the same as saying the first juggler can make his division. Ball (p. 475) simply states the answer with no explanation or source. Since this is likely to have occurred in the first edition of 1890, this is now probably the earliest known modern version. This makes the possible connection with India more feasible as Ball's *A Short Account of the History of Mathematics* appeared in 1888 and was well-known in the English-speaking world, so Rangāchārya could well have corresponded with Ball.[d]

Edward Wakeling [35] discusses the problem "Four brothers and a monkey." Wakeling gives the solution 765 and says there are other solutions, citing 2813 and 5885, i.e. $765 + 2048k$, but the general solution is actually $765 + 1024k$.

Now the question of the date of Carroll's version is difficult to pin down. This problem was found in a collection of papers of Professor Bartholomew Price (1818–1898). He was at Pembroke College, becoming the Master, adjacent to Carroll's Christ Church. He had tutored Carroll (1833–1898) and they were close friends until their deaths. Price had a mass of papers sent to him by Carroll, including this problem, but few of the papers are dated and they are simply unordered loose sheets, so there is no way to date the sheets such as that containing this problem. Wakeling says much of this material was intended for Carroll's projected book on puzzles.

The fact that both Ball and Carroll knew of the problem leads one to suspect that it may have been known to others in the 1890s. After the appearance of *Alice* in 1865, Carroll's reputation was immense and he could easily have been sent the problem from anywhere in the world. Wakeling kindly sent me copies of three items of the Carroll/Price material. First is Carroll's solution of the problem, which is typewritten, "probably using Dodgson's Hammond

[d]However, I have examined the Ball material at Trinity College Cambridge and it is clear that he (or his heirs) disposed of his correspondence, so this conjecture cannot be verified.

The number of nuts that each of the brothers has to divide must be a multiple of 4, and the residue after taking one fourth a multiple of 3 . Therefore the number the third brother has to divide must be a multiple of 4 which with one added to it is a multiple of 3; or (which is the same thing) a multiple of 4 which, divided by 3, leaves 2 over . 8 is such a ~~number~~multiple and by successive additions of 12 (3 times 4, or 4 times 3) we get all the multiples of 4 that fulfil the same condition. The number, therefore, to be divided by the third brother, must be either 2,5,8,11, or 14 &c:times 4; and when 1 is added to it, it is equal to 3,7,11, or 15 &c:times 3 . This is the residue after the second brother's division, hence the number he had to dividemust have been 3,7,11, or 15 &c:times 4. But by applying the previous reasoning, we know that the number the second has to divide must be 2,5,8,11,or 14&c:times4 . Hence the least number that will do is the smallest that occurs in both series, i.e. 11 times 4 = 44; adding 1 gives 45, the residue of the first division, or 60 before the first division; and adding one for the nut abstracted = 61. (Answer.)

It is also obvious that the other answers are obtained by successive additions of 64 to the first answer, as 125, 189, 253&c:

CLD's typed solution (c.1888)

Nash Mills,
Hemel Hempstead.

Oct 15 1888

Dear Prof Price

Many thanks for your solution as to the 4 Brothers & the Monkey. I had somehow got as high as 1789. I think that my hypothesis and justification will justify what has been said about them

Yours sincerely
John Evans

On trying your solution I find that Brother No 4 finds 105 nuts & after he gives one to the monkey leaving 104. Of these he takes 26 leaving 78 which is not divisible by 4 and therefore does not fulfil the conditions. Nor does I think 509 nor 765

Figure 13.5. Carroll's solution(?), c .1890, from Wakeling; and Evans' letter of 15 October 1888, from Wakeling.

typewriter, purchased in 1888." This is in Figure 13.5. This solution is grossly erroneous — he only takes three stages and obtains the answer $61 + 64k$.

Most importantly, Wakeling sent a note from John (later Sir John) Evans to Price, dated 15 Oct 1888, thanking Price for his solution of the problem and saying that his attempt had gotten to a value of 1789 (which is a correct solution!), see Figure 13.5. Evans then adds that he cannot make Price's solution work. Price must have given 253, but after the fourth brother, there remain 78 which is not divisible by four (nor is it one more than a multiple of four). Evans then says that $509 (= 253 + 256)$ and $765 (= 253 + 512)$ also fail, "I think." Wakeling also sent the statement, only, of the problem, in Evans' handwriting, headed "Four Brothers & the Family Monkey" — this differs from the version in Wakeling. Though this is a moderately messy problem, it is depressing to discover that three competent mathematicians were unable to get the correct solution and failed to check the solutions that they had obtained! However, we now know that the problem was in circulation in 1888, and the fact that wrong answers were being obtained shows that the problem was new at that time.

The mathematician always feels that a question should have an answer and that the answer will eventually be found. But problems like the Coconuts problem show that historical questions are often unanswerable and probably never will be answered or will only be answered by serendipity — e.g. if Pearson's correspondence turns up in a descendant's attic or if an earlier US publication turns up.

13.3 A General Solution

Before approaching the general solution, let us examine what Mahāvīrā does. He simply gives a rule for the answer, but it is all expressed in words and not easy to understand. The translator notes that it does the second problem by computing $x_1 = (2x_0 - 2)/3$, then $x_2 = (4x_0 - 10)/9$ and $x_3/3 = (8x_0 - 38)/81$ and then requiring that all of these be integers. However, the translator only gives the smallest positive answers: 11 for the first problem and 25 for the second problem. One can view this as solving $8x \equiv 38 \pmod{81}$, which is a form of the Chinese remainder theorem which was well known in India by the 7th century.

The mathematics of these problems is actually fairly simple, but takes a bit of careful work. There are a number of stages where the amount is changed in a regular manner. Let x_i be the amount present after the ith stage. In most of the problems, we can express the division by

$$x_{i+1} = \alpha x_i + \beta.$$

For example, for the original coconuts problem, $x_{i+1} = \frac{4}{5}(x_i - 1)$, so $\alpha = -\beta = \frac{4}{5}$, corresponding to the monkey getting one before the division takes place. But for the dell'Abbaco version, we have $x_{i+1} = \frac{1}{2}x_i - 2$. This is the more usual situation for determinate problems, where the subtraction takes place after the division. Euler seems to be the only example of the other case for a determinate problem, though Bhāskara has subtraction both before and after the division. Scot Morris [25], pointed out that one can have a coconuts problem with subtraction after division, i.e. each sailor takes his fifth and then gives one from the remaining pile to the monkey. For the case of 5 sailors, this would be $x_{i+1} = \frac{4}{5}x_i - 1$, with $\alpha = \frac{4}{5}, \beta = -1$.

In many of the determinate problems, both α and β vary with i. Since determinate problems pose few difficulties, we shall now concentrate on the coconuts problems — the reader can easily adapt the methods to the determinate case. Suppose we have n sailors in a coconuts problem. Then we have seen two forms:

$$\text{Form 0. } x_{i+1} = \frac{n-1}{n}x_i - 1.$$

$$\text{Form 1. } x_{i+1} = \frac{n-1}{n}(x_i - 1).$$

In both cases, $i = 0, 1, \ldots, n-1$. The notation points out that in Form F, each $x_i \equiv F \pmod{n}$ in order for x_{i+1} to be an integer. One can easily extend this to allow subtracting c coconuts at each stage, giving us the following.

$$\text{Form 0}c\text{. } x_{i+1} = \frac{n-1}{n}x_i - c.$$

$$\text{Form 1}c\text{. } x_{i+1} = \frac{n-1}{n}(x_i - c).$$

The previous cases arise when $c = 1$.

Before attacking the solution of these equations, we need to examine the ending process. The versions of Clark and Pearson have $x_n \equiv 1 \pmod{n}$ and then the monkey gets an extra coconut at the end. But the original Coconuts version of Williams has $x_n \equiv 0 \pmod{n}$, so the monkey does not get a final coconut. This is perhaps the most natural way to have an ending for a Form 0 problem, but we can impose the final condition $x_n \equiv E \pmod{n}$, with the monkey getting E from the final pile. Normally, we would have $E = c$ or $E = 0$. We shall denote this by saying that the problem has "Ending E."

Now we consider the general process of solving

$$x_{i+1} = \alpha x_i + \beta. \tag{13.1}$$

This is an example of a difference equation or a recurrence relation. There are several standard methods which can be applied here.

Some knowledge of these methods tells us that there will be a solution of the form $x_i = A\alpha^i + B$ and one can simply determine A and B. (This is the "method of undetermined coefficients.") However, this method is a bit magical, so let me show a way of obtaining the solution. Suppose we shift the values of the sequence by B, so we let $y_i = x_i - B$. Then $y_{i+1} = x_{i+1} - B = \alpha x_i + \beta - B = \alpha(y_i + B) + \beta - B = \alpha y_i + (\alpha - 1)B + \beta$. So if we chose $B = \beta/(1-\alpha)$, then our equation reduces to the "homogeneous" problem:

$$y_{i+1} = \alpha y_i. \tag{13.2}$$

This obviously has the solution

$$y_i = \alpha^i y_0, \tag{13.3}$$

so that the solution of Eq. (13.1) is

$$x_i = \alpha^i (x_0 - B) + B, \quad \text{where} \quad B = \frac{\beta}{1-\alpha}.$$

Note that Eq. (13.1) has the constant solution $x_i = B$. We can also say that B is the fixed point of the equation $B = \alpha B + \beta$, which is an easy way to determine B. In our two basic forms, the parameters are as follows.

$$\text{Form 0}c.\ \alpha = \frac{n-1}{n}; \quad \beta = -c; \quad B = -nc.$$

$$\text{Form 1}c.\ \alpha = \frac{n-1}{n}; \quad \beta = -\frac{n-1}{n}c; \quad B = -(n-1)c.$$

So far, so good. But now we realize that there is more to consider. All our problems require that all x_i be integers. Since B is integral, this is equivalent to asking for all y_i to be integers. From Eq. (13.3), we have $n^i y_i = (n-1)^i y_0$. Since n and $n-1$ are relatively prime, this requires n^i to divide y_0. This must hold for $i = 0, 1, \ldots, n$ and the last case implies the others. If $y_0 = An^n$, then $y_i = An^{n-i}(n-1)^i$ and we have

$$y_0 = An^n \quad \text{and} \quad y_n = A(n-1)^n \tag{13.4}$$

and

$$x_0 = An^n + B \quad \text{and} \quad x_n = A(n-1)^n + B. \tag{13.5}$$

This gives us all the integral solutions for our standard forms, but we now have to consider the endings. For Ending E, we want $x_n \equiv E \pmod{n}$, i.e. $A(n-1)^n + B \equiv (-1)^n A + B \equiv E \pmod{n}$, which gives us $A \equiv (-1)^n(E-B) \pmod{n}$. If we write this as $A = (-1)^n(E-B) + Cn$, we have $x_0 = (-n)^n(E-B) + Cn^{n+1} + B$, so that

$$x_0 \equiv (-n)^n(E-B) + B \pmod{n^{n+1}}. \tag{13.6}$$

Using the parameters given above, the solution for $x_0 \pmod{n^{n+1}}$ is tabulated below.

Form	Ending 0	Ending 1	Ending c
$0c$	$-nc$	$(-n)^n - nc$	$(-n)^n c - nc$
$1c$	$-(-n)^n c - (n-1)c$	$(-n)^n(1-c) - (n-1)c$	$-(n-1)c$

The smallest positive solutions depend a bit on c, so we shall just consider the case $c = 1$ in the Forms and just the Endings 0 and 1, which are the most natural cases. The solutions $(\bmod\ n^{n+1})$ are given below for these cases.

Form	Ending 0	Ending 1
0	$-n$	$(-n)^n - n$
1	$-(-n)^n - (n-1)$	$-(n-1)$

The values of $-n$ and $-(n-1)$ are the simple solutions with negative coconuts, as noted by Anning [31] in 1912. The smallest positive solutions depend on the parity of n and are given below.

Form	Ending 0	Ending 1
0	$n^{n+1} - n$	Odd: $n^{n+1} - n^n - n$ Even: $n^n - n$
1	Odd: $n^n - (n-1)$ Even: $n^{n+1} - n^n - (n-1)$	$n^{n+1} - (n-1)$

This would seem to completely solve our problems, but there are two unexpected features after this point.

After first reaching this stage, we computed answers for all the problems. If we apply these formulas to Mahāvīrā's first problem, we have Form 1, Ending 0, $n = 2$ and these give us $x_0 \equiv 3 \pmod{8}$.

The least positive solution is $x_0 = 3$. But this gives us $x_1 = 1$, $x_2 = 0$ and the last division is trivial. From Eq. (13.5), we see that $x_n = 0$ gives us $A(n-1)^n = -B$. In Form 0, this gives us $A(n-1)^n = n$, which can only hold if $n = A = 0$ or $n = A = 2$, and the latter is the case when $x_0 = 6$. In Form 1, we get $A(n-1)^n = n-1$, which can only hold if $n = A = 1$ or $n = 2, A = 1$, and the latter is the case when $x_0 = 3$. Thus this feature, of getting to zero, is only significant for $n = 2$.

Now let us compute the sequence of values x_i for some typical cases. Below are the values for $n = 4$ and $n = 5$.

$n = 4$

(Form, Ending)=		(0, 0)	(0, 1)	(1, 0)	(1, 1)
$i=$	0	1020	252	765	1021
	1	764	188	573	765
	2	572	140	429	573
	3	428	104	321	429
	4	320	77	240	321
$(x_4 - E)/4$		80	19	60	80

$n = 5$

(Form, Ending)=		(0, 0)	(0, 1)	(1, 0)	(1, 1)
$i=$	0	15620	12495	3121	15621
	1	12495	9995	2496	12496
	2	9995	7995	1996	9996
	3	7995	6395	1596	7996
	4	6395	5115	1276	6396
	5	5115	4091	1020	5116
$(x_5 - E)/5$		1023	818	204	1023

When I computed these values, the patterns jumped out at me. The similarity of the first and last columns is easy to see. Consider the process of (subtract 1, then delete $\frac{1}{5}$) five times to leave $1 + 5d$. If we take away the first 1 and shift the stages, we are doing the process of (delete $\frac{1}{5}$, then subtract 1) five times to leave $5d$, so the values in the (0,0) problem are just one less than the values in the (1,1) problem and the final division gives the same amount in both cases.

But the relation between the (0,0) and (0,1) problems for n odd and the (1,0) and (1,1) problems for n even is not so easy to see. Indeed, it cannot be as simple as the previous relation as it depends on the parity of n. One can simply check that starting with x_0 in one case gives x_1 which is the x_0 of the other case. Perhaps it is a little clearer if one determines the values of c and A and finds that in the case n odd, the (0,0) problem has $y_0 = nn^n$, while the (0,1) problem has $y_0 = (n-1)n^n$. From the formula for y_i, it is then clear that one sequence is the same as the other shifted by one place.

Tony Barnard, of King's College London, kindly provided an alternate formulation of the solution when $c = 1$. In this case, he observes first that $B = -n + F$, where $F = 0$ or 1 is the form of division. Hence Eq. (13.5) can be rewritten as

$$x_0 = An^n - n + F \quad \text{and} \quad x_n = A(n-1)^n - n + F, \qquad (13.7)$$

and, in general, $x_i = A(n-1)^i n^{n-1} - n + F$. A is determined as the least positive integer such that $x_n \equiv E \pmod{n}$, i.e. $(-1)^n A \equiv E - F \pmod{n}$. Then one only needs to have a table of values of A, as follows.

F	E	parity of n	A
0	1	even	1
1	0	odd	1
0	1	odd	$n-1$
1	0	even	$n-1$
0	0	–	n
1	1	–	n

Barnard's version clearly shows that going from (0,0) to (0,1) in the odd case changes A from n to $n-1$ which corresponds to shifting the sequence of values by one place. However, it is still unclear how to rephrase the versions so that this connection is intuitive in the same way that the connection of the (0,0) and (1,1) versions is.

As a parting problem, let me describe a variation found in Fibonacci [15].

> A dying man has several sons and he leaves them his estate as follows. The first son gets one bezant and $\frac{1}{7}$ of the remainder. The second son gets 2 bezants and $\frac{1}{7}$ of what then remains, the third son gets 3 bezants and $\frac{1}{7}$ of the remainder, etc. Afterward, the sons compare and discover they have all received the same amount!! How many sons were there and what was the old man's estate?

Chuquet [11] gives many versions of this, some of which have non-integral numbers of children!

13.4 Other Variations

Leroy F. Meyers referred me to [29] which describes and solves another version of the problem from Birkhoff and MacLane [8]. This has the usual five men, a monkey and some coconuts, i.e. Form 1, $c = 1, n = 5$, but they specify no ending — they simply want x_n to be integral. Pedoe *et al.* [29] consider the problem with general c and with any number of steps m. They observe that the usual Ending c version is the same as the [8] version but with $m = n + 1$ instead of $m = n$.

Looking at Eq. (13.5) or (13.7) above and recalling that B and F are integral, we see that x_n is integral for any integral A and the least positive solution is given by $x_0 = n^n + B$ in general. Extending Tony Barnard's comment, we see that $B = -(n - F)c$, so $x_0 = n^n - nc + Fc$ in general. We then have $x_n = (n-1)^n - nc + FC \equiv (-1)^n + Fc \pmod{n}$. This is not congruent to 0 or 1 or $c \pmod{n}$ except in some special cases. That is, the solutions in [8] are generally different than the solutions we have considered above, but we shall examine the cases where there is some connection.

When n is odd, we have $x_n \equiv -1 + Fc \pmod{n}$. For Form 1 and $c = 1$, this gives an Ending 0 solution. For example, in [8] $c = 1$, $m = 5$, so it has the same solution as our $F, E = 1, 0$ problem. Taking Form 1, $c = 2$ would give an Ending 1 solution.

When n is even, we have $x_n \equiv 1 + Fc \pmod n$. For Form 0, this gives an Ending 1 solution. For Form 1, $c = 1$, this gives an Ending 2 solution, which is simply one greater than the solution of our $F, E = 0, 1$ problem, and this can be understood by the same sort of argument as above. For example, if we consider the problem in [8], with $n = 4$, then the least positive solution is $x_0 = 253$.

Kircher [21] generalizes the problem considerably by taking any number of sailors, any number of divisions and allowing the ith division to discard a variable amount V_i before taking away $1/n$ of the rest, even allowing negative V_i, e.g. if the monkey is adding coconuts to the pile! Sadly, his basic recurrence equations are misprinted. He solves this form of the problem by use of difference calculus, but the generalization and its solution seem do not seem close enough to the classical problem to discuss here.

One can extend the problem a bit by having k divisions of the original pile. The analysis leading to Eq. (13.4) is easily extended to show $y_0 = An^k$, and one can carry this through the remaining steps if one wishes.

In 1951, Norman Anning [4] studied the problem with three men, ending 1, getting 79 coconuts, etc. He then looked at the general case with n men and s monkeys, and pointed out that loaning $(n-1)s$ coconuts simplifies the calculations, i.e. it makes the recurrence homogeneous as above.

In 1954 [5] there was a variation that had three men and cigarettes guarded by a Boy Scout. Two are given to the scout at each division and at the end, so this is Form 1c, Ending c, with $c = 2$. Solution observes that $-(n-1)c$ is a fixed point and the solution is $-(n-1)c \pmod{n^{n+1}}$, as seen in the table after Eq. (13.6) above, but the solution doesn't give any proof.

13.5 Solutions and Some Comments

[36] 15621
[15] (p. 278) 382
[26, 27] 295
[19] 42
[15] (pp. 316–318) 59
[3] $21\frac{7}{8}$
[10] (pp. 79–80) $30468\frac{84876}{371293}$
[10] (p. 69) $10\frac{15}{16}$
[30] (no. 24) $16\frac{5}{8}$
[30] (no. 28) 15
[10] (pp. 69–70) $\frac{6}{5}$
[7] 56
[15] (pp. 266–267) This leads to the general expression $x_i = 14 - 2^i$.
[11] (prob. 95) The recurrence is $x_i = \frac{4}{3}x_{i-1} + i$. The expression for x_i can be got by brute force. But if one tries for a solution in this particular problem by setting $y_i = x_i + a + b_i$, then one easily gets $a = 12, b = 3$, and so $x_n + 3n + 12 = (\frac{4}{3})^n(x_0 + 12)$.
[14] 1480
[24] There were 81 potatoes originally and the men now have 27, 18, 12, so the remaining 24 should be given out as 0, 9, 15.
[31] 79
[28] (no. 29) 79
[28] (no. 94) 1021
[12] 79
[13] $1048575 = 2^{20} - 1$. The general solution is $2^{20k} - 1$.
[23] (pp. 124–125) 3 is the answer given by our formula, but this causes $x_n = 0$, which seems a bit odd, and Mahāvīrā has taken the next case, namely 11, as the solution.
[23] (p. 125) 25
[15] (pp. 279–281) 6 sons, 36 bezants
[2] (prob. 71) $123\frac{4}{4}$
[9] 500

Bibliography

[1] DELL'ABBACO-ARITMETICA Prob. 47, p. 44 with plate on p. 45. See also prob. 71, pp. 65–67 with plate on p. 66.

[2] DELL'ABBACO-TUTTA 438 is a reproduction showing a version of our problem.

[3] Anania Schirakatzi (= Ananias of Shirak). *Arithmetical problems.* Armenian MS, c. 640. Translated by: P. Sahak Kokian "Des Anania von Schirak arithmetische Aufgaben," *Zeitschrift für d. Deutschösterr. Gymnasien*, 69 (1919) 112–117. Prob. 19.

[4] N. Anning. "Monkeys and coconuts." *Mathematics Teacher*, 44 (December 1951) 560–562.

[5] Anonymous. "The problems drive." *Eureka*, 17 (October 1954) 8–9 and 16–17. No. 2.

[6] W. W. Rouse Ball. *Elementary Algebra*, 2nd ed. Cambridge Univ. Press, 1897, 260. Example 12.

[7] BHĀSKARA II. 196–197.

[8] G. Birkhoff and S. Mac Lane. "Prob. 11." in *A Survey of Modern Algebra.* Macmillan (1941), p. 26 (p. 28 in the 4th ed. of 1977).

[9] CALANDRI.

[10] CHIU. Chap. VI, prob. 27–28, pp. 69–70; Chap. VII, prob. 20, pp. 79–80.

[11] CHUQUET. Prob. 95. Also Prob. 129–141.

[12] S. E. Clark. *Mental Nuts. Can you Crack 'em? A book of 100 Catch or Trick Problems.* Waltham Watches, 1897 (rev. ed. 1904). [A small advertising booklet.]

[13] H. E. Dudeney. "Problem 522." *Weekly Dispatch* (8 and 22 November 1903) both p. 10.

[14] EULER Questions for practice, no. 9, p. 204.

[15] FIBONACCI Chap. 12, part 6, 258–267, 278–281, 313–318 & 329. *De viagiorum propositionibus, atque eorum similium* is devoted to such problems.

[16] G. Flegg, C. Hay and B. Moss. *Nicolas Chuquet, Renaissance Mathematician. A study with extensive translation of Chuquet's mathematical manuscript completed in 1484.* Reidel, 1985.

[17] M. Gardner. "The Monkey and the Coconuts." *The Second Scientific American Book of Mathematical Puzzles and Diversions*, Simon & Schuster, 1961, 104–111. [UK version: *More Mathematical Puzzles and Diversions*; Bell, London, 1963; Penguin, 1966.]

[18] H. Hermelink. "Arabische Unterhaltungsmathematik als Spiegel Jahrtausendealter Kulturbeziehungen zwischen Ost und West."

Janus 65 (1978) 105–117, with English summary. (In English: "Arabic recreational mathematics as a mirror of age-old cultural relations between Eastern and Western civilizations." in *Proc Intl Symp History of Arabic Science—Vol. Two: Papers in European Languages.* Inst for the History of Arabic Science, Aleppo, 1978, 44–52.

[19] JACKSON. 17 & 74.

[20] G. R. Kaye. "A brief bibliography of Hindu mathematics." *J. Asiatic Soc. Bengal* (*NS*), 7, 10 (November 1911) 679–686.

[21] R. B. Kircher. "The generalized coconut problem." *Amer. Math. Monthly*, 67, 6 (June/July 1960) 516–519.

[22] LUCAS. p. 184.

[23] MAHĀVĪRĀ(CHĀRYA). pp. 124–125.

[24] L. Mittenzwey. "Prob. 138" in *Mathematische Kurzweil.* (7th ed.) 1879 (also Julius Klinkhardt, 1918), 26 & 76–77.

[25] S. Morris. "The monkey and the coconuts." *The Next Book of Omni Games.* Plume (New American Library), 1988, 30–31 & 182–183.

[26] OZANAM. Prob. 28, question 1, 1725, 211–212.

[27] OZANAM-MONTUCLA. Prob. 15, 207–209.

[28] PEARSON. Part II. No. 29: "The men, the monkey, and the mangoes," 119 & 197. No. 94: "One for the parrot," 133–134 & 210.

[29] D. Pedoe, Timothy Shima and Gali Salvatore. "Of coconuts and integrity." *Crux Mathematicorum*, 4, 7 (August/September 1978) 182–185.

[30] THE RHIND PAPYRUS. 37–40 & 54–55.

[31] "Problem 288." *School Science and Mathematics* 12 (1912) 235, 520–521. (Nelson L. Roray, proposer; A. M. Harding, Norman Anning and the proposer, solvers.)

[32] D. E. Smith. "The Ganita-Sara-Sangraha of Mahāvīrāchārya." *Bibliotheca Mathematica*, 3 (1908/09) 106–110.

[33] ṬABARĪ. 127, no. 44.

[34] "Problem 3242." *Amer. Math. Monthly*, 34 (1927) 98; 35 (1928) 47–48. (R. S. Underwood, proposer; R. E. Moritz, solver.)

[35] Edward Wakeling. *Lewis Carroll's Games and Puzzles.* Dover, 1992, problem 17, 21 & 68. With the Lewis Carroll Birthplace Trust.

[36] Ben Ames Williams. "Coconuts." *Saturday Evening Post* (9 October 1926) 10, 11,186,188. (also in Clifton Fadiman, ed.; *The Mathematical Magpie*; Simon & Schuster, 1962, 196–214.)

Chapter 14

Two River Crossing Problems

Ian Pressman and David Singmaster

River crossing problems first appear in the *Propositiones ad Acuendos Juvenes*, attributed to Alcuin of York (c. 732–804) — see Chap. 4. Alcuin gives four such problems, of which two are essentially identical. The first (Prop. 17) is the problem of the three jealous husbands. The second (Prop. 18) is the problem of the wolf, the goat and the cabbage. The third (Prop. 19) is the problem of the two adults and the two children, where the children weigh half as much as the adults. The fourth (Prop. 20) is a rewording of the previous problem.[a] We only focus on the first of these.

Alcuin's phrasing of the problem is much more direct than later versions, so we repeat the translation, from Chapter 4.

> Three men, each with a sister, needed to cross a river. Each one of them coveted the sister of another. At the river, they found only a small boat, in which only two of them could cross at a time. How did they cross the river, without any of the women being defiled by the men?

The Latin is a bit ambiguous about who covets whom, but the usual interpretation is that we cannot allow any sister to be with another man without the protection of her brother. Alcuin gives a correct solution in 11 crossings without anything untoward happening.

---

[a]This chapter appeared in somewhat revised and shortened form in *Mathematical Gazette*. 73 (464) June 1989, 73–81. This is joint work with Ian Pressman. I proposed it among other projects for final year B.Sc. students and Ian decided to work on it.

These problems apparently were widely disseminated as they appear in many of the problem collections of the next centuries, with the notable exception of Fibonacci. Ahrens [1] gives a verse solution which may be as old as 10th century and many later references to the wolf, goat and cabbage problem which we have not seen. The 13th century *Annales Stadenses* compiled by Abbot Albert von Stade calls the wolf, goat and cabbage problem a problem for children. Further he gives a mnemonic; see Problem 6 in Chapter 5 for a full discussion. The c. 1370 *Columbia Algorism* [9] gives a simple but delightful illustration, Figure 14.1.

The first person to consider more couples appears to be Luca Pacioli, in his unpublished manuscript *De Viribus Quantitatis* [18], where he asserts that 4 or 5 couples requires a three person boat. However, Tartaglia, in 1556 [21] asserts that 4 couples can cross in a two person boat and gives a sketchy solution. Bachet [6] pointed out Tartaglia's error and showed that 4 couples cannot cross in a two person boat. Tartaglia's error is worth examining since it clarifies what is considered acceptable. Tartaglia transfers the four women to the other side, then sends one back. She remains and two men other than her husband get in the boat. When they reach the far side, the unaccompanied woman gets in the boat and returns with it. This is the point where Bachet and other authors object — when the men reach the far side, there are two men and three women there, which is not permitted by the usual interpretation of the jealousy conditions.

Figure 14.1. From the *Columbia Algorism*, c. 1370.

14.1 De Fontenay's Generalization

In the late 19th century, Lucas [15] stimulated a revival of interest in recreational problems among French mathematicians and students.

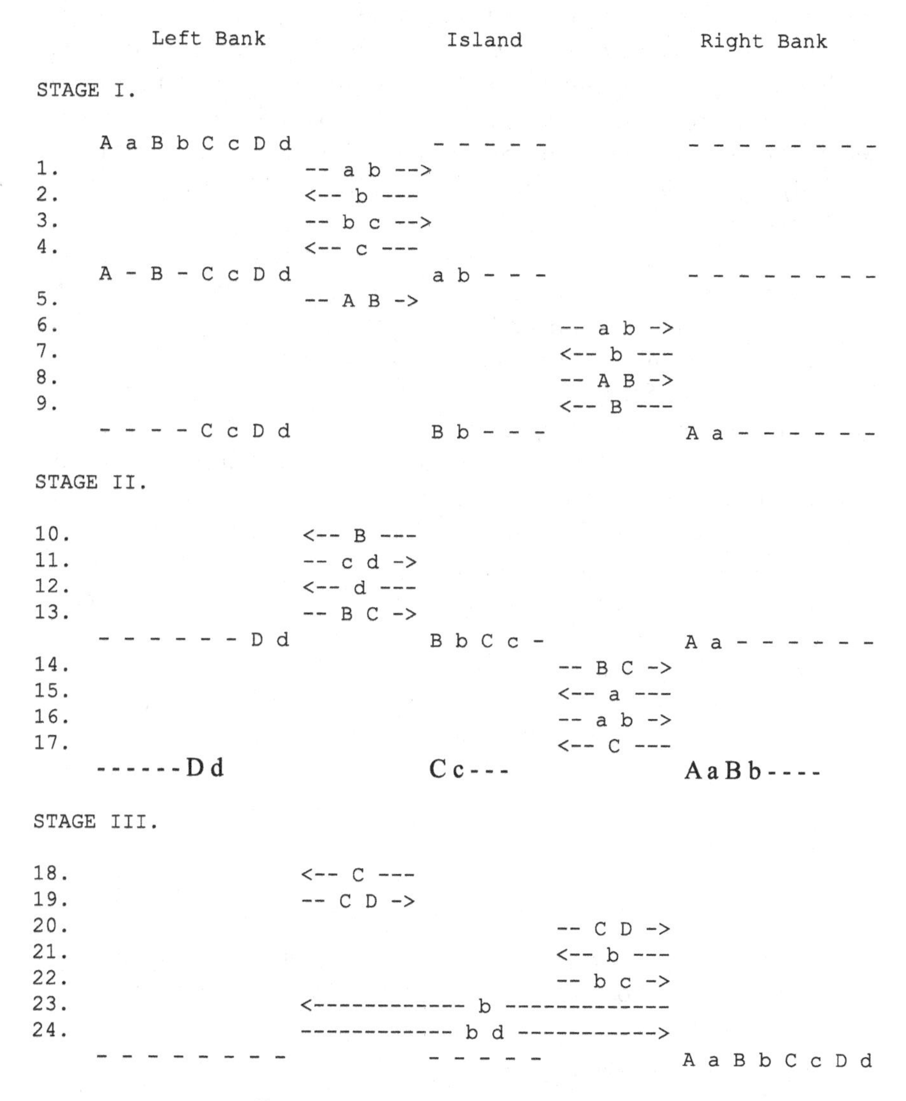

Figure 14.2. De Fontenay's solution.

In 1879, a young student, M. Cadet De Fontenay,[b] observed that 4 or more couples could cross if there were an island in the river. He gave a solution for 4 couples in 26 crossings, which extends to n couples in $8n - 8$ crossings. Below, we show De Fontenay's solution in the usual and convenient diagrammatic form, though we do not show each of the dispositions of people. The couples are denoted `A a`, with `A` the husband, etc. The solution has three stages.

The stages take 9, 8, 7 moves. Noting that the difference between the beginning and the end of Stage II is simply equivalent to moving one couple across the river, we see that n couples can be moved across the river by repeating Stage II $n - 3$ times, giving a total of $8(n - 3) + 16 = 8n - 8$ crossings. (Lucas says $4n - 4$ voyages, which must mean two-way trips.)

When we examined the above solution, we were greatly perplexed. Because the position at the beginning of Stage III is symmetric to the position at the end of Stage I, it seemed that these stages should take the same number of crossings! Indeed, a little work shows that both Stage I and Stage II can easily be reduced to 7 crossings, by using bank to bank crossings as in moves 23 and 24, and one can go further. Lucas' presentation is not as detailed as the above so one cannot tell how he intended to accomplish our moves 23 and 24.

It appears that De Fontenay and/or Lucas did not permit bank to bank crossings. The count of moves implies moves 23 and 24 as above, but the number given may be erroneous. Certainly, Dudeney later did permit bank to bank moves and this easily permits even greater reductions in the number of moves, as we shall see below.

If bank to bank moves are prohibited, then the last two moves above must be replaced by the following.

```
23'.                                     <-- b ---
24'.                <-- b ---
25'                 -- b d ->
26'.                                     -- b d ->
     - - - - - - - - -      - - - - -              A a B b C c D d
```

We now see that this gives a solution for n couples in $8n - 6$ crossings, with no bank to bank crossings. Pressman's computer program

---

[b]Regarding the spelling of De Fontenay/De Fonteney. Lucas and Ahrens give De Fontenay, but Ball and his French translator give De Fonteney. I have discovered there is a famous Abbey of Fontenay, so this form seems more likely.

showed that this was the minimum number of moves for $n = 4$ and later for $n = 5, 6, 7, 8$, which led us to the following simple result.

Theorem 1. *The above method is optimal, that is, the minimal number of crossings to ferry n couples across a river with an island, using a two person boat and with no bank to bank crossings, is $8n-6$ moves for $n \geq 1$.*

Proof. Consider any method of ferrying $2n$ people across the river without bank to bank crossings. Each departure and return from the left bank results in at most one person being taken from the left bank to the island, except that two persons can go on the last departure. Hence there must be at least $2(2n-2)+1 = 4n-3$ crossings involved. The same holds for the transits from the island to the right bank, so overall there must be at least $8n - 6$ crossings in any ferrying of $2n$ people. As we have seen, the (amended) method of De Fontenay

```
STAGE I.

    A a B b C c D d          - - - - -         - - - - - - - -
1.               ------------ a b ----------->
2.               <------------ b -------------
3.               -- b c ->
4.               <-- c ---
5.               ------------ A B ----------->
6.               <------------ B -------------
    - - B - C c D d          b - - - -         A a - - - - - -

STAGE II.

7.               -- c d ->
8.               <-- d ---
9.               ------------ B C ----------->
10.                                <-- a ---
11.                                -- a c ->
12.              <------------ B -------------
    - - B - - - D d          b - - - -         A a - - C c - -

STAGE III.

13.              ------------ B D ----------->
14.                                <-- c ---
15.                                -- b c ->
16.              <------------ c -------------
17.              ------------ c d ----------->
    - - - - - - - -          - - - - -         A a B b C c D d
```

Figure 14.3. Dudeney's solution.

manages to ferry jealous couples in this number of moves, so the method must be optimal. □

There is a lengthy discussion of the problem by Lucas [14]. In "Prob. XXXVI: La traversée des trois ménages," Lucas gives Bachet's 1612 reasoning for the essentially unique solution for 3 couples. He later considers larger boats, then Tartaglia's error and Bachet's notice of it and gives an easy proof that 4 couples cannot be done with a 2 person boat. "Prob. XXXVIII: La station dans une île," has 4 couples, 2 person boat, with an island and he gives De Fontenay's solution in 24 crossings. "Note II: Sur les traversées," gives Tarry's polygamous version and solutions in various cases. For the ordinary case, he finds a solution for 4 couples in 21 moves, using the basic ferrying technique that we found to be optimal, but not getting the beginning and ending optimized. He says this gives a solution for n couples in $4n + 5$ crossings. He then considers the case of $n - 1$ couples and a ménage with m wives and finds a solution in $8n + 2m + 7$ crossings.

14.2 Dudeney's Solution

A later author who worked on De Fontenay's problem is Dudeney [10], who used bank to bank crossings and obtained a solution in 17 crossings for 4 couples. "It can be done in seventeen passages from land to land (though French mathematicians have declared in their books that in such circumstances twenty-four are needed), and it cannot be done in fewer." We present his solution below. Again, it has three stages.

Ball [7] discusses these problems under the heading of "Ferry Boat Problems." In the early editions, he states De Fontenay's problem and solution (spelling the name as De Fonteney). From the 9th edition of 1920 (and perhaps earlier — we haven't seen every edition), he says: "It would, however, seem that if n is greater than 3, we need not require more than $6n - 7$ passages from land to land.†" The footnote (†) cites Dudeney. Examining Dudeney's solution, we see that Stage II has the effect of moving one couple across the river in 6 crossings. Repeating Stage II $n - 3$ times, we have a solution for n couples in $6n - 7$ crossings as asserted by Ball. In the 12th ed. of 1974, Coxeter has amended Ball's $8n - 8$ of De Fontenay's solution to the more correct $8n - 6$, but he still shows the same solution given by

Lucas and says the stages take 9, 8, 7 steps! Ball first states that the problem can be solved in $6n - 7$ crossings in the 6th ed. of 1914, but he gives no reference for it. He first cites Dudeney for the solution in $6n - 7$ crossings in the 8th ed. of 1919, pp. 71–73.

14.3 Improved Solutions

An initial observation about the problem is that there are $2n + 1$ objects (counting the boat) which can be in any one of three places, so there are 3^{2n+1} distinct positions that might be considered. For

THE MISSIONARIES' AND CANNIBALS' PUZZLE.

Three Missionaries, accompanied by three Cannibals, arrive at a river in Africa, and desire to cross. There is but one boat there, and that will hold but two at a time. All the Missionaries and one of the Cannibals, the shorter one, can row. The problem is: How to get all across the river in the boat without at any time leaving one or two Missionaries with a greater number of Cannibals. If there were two Cannibals left on one shore with one Missionary, or three Cannibals with two Missionaries, the Missionaries would surely be eaten. It should be remembered that only one of the Cannibals can row.

TO TRY THIS PUZZLE.

Take six matches, break off the heads of three, to represent the Cannibals. Make one of these shorter, to represent the little Cannibal who can row. Draw an imaginary line to represent the river, and, placing all on one side, commence. If desired, a small piece of paper can represent the boat, though this is not necessary.

The solution of every puzzle in this book sent free, on receipt of an Allcock Plaster or Brandreth Pill wrapper.

Figure 14.4. From *The Brandreth Puzzle Book*, 1896 [8].

$n = 4$, this gives $3^9 = 19683$ positions. Obviously many of these violate the jealousy restraints and many are clearly equivalent. Also, the boat cannot be somewhere where there are no people!

We wanted to do a computer search. Our first idea was to generate all the inequivalent permissible distributions of people and boat, then generate edges which join two distributions which differ by a single crossing and then use a standard graph-theoretic algorithm to find the shortest path which moves everyone to the other bank. Pressman suggested a direct breadth-first search which would generate the positions as it searched. This led us to ask how many inequivalent permissible distributions there were. We recognized that an equivalence class of permissible positions is completely determined by the number of men and women at each site. Let these numbers be M_i, W_i for $i = L, I, R$. Then we must have $M_L+M_I+M_R = W_L+W_I+W_R = n$ and the jealousy conditions are that $M_i \geq W_i$ or $M_i = 0$, for each i. A simple computer count of such distributions revealed rather small numbers and then we were able to evaluate the number exactly by somewhat tedious enumeration of cases. The result is that there are $\frac{3}{2}n(n^2 + 10n - 1)$ inequivalent permissible positions. For $n = 4$, this gives 330 such positions, which is much less than the 19683 initially considered. The fact that the number was only a cubic in n made us feel that almost any searching method would find optimal solutions.

Pressman implemented this suggestion, first for $n = 4$ without bank to bank crossings, then with an option to permit bank to bank crossings and finally for $n = 4, 5, 6, 7, 8$. We were delighted to find that the computer found a solution for $n = 4$ in 16 crossings. This involves an ingenious step which Dudeney apparently failed to consider. In trying to understand this solution, we found a solution for n couples that used $4n + 1$ crossings and then the computer showed that this was minimal for $n = 5, 6, 7, 8$. We have now proven that this is minimal for all $n > 4$ and the proof is presented below.

Observing that Stage II moves a couple across the river, we see that it can be repeated $n-2$ times to move n couples across the river in $4n + 1$ moves. Stage II′ shows such a repetition. The simplicity of this method startled us and we are quite surprised that it was not found sooner. One simply parks two wives on the island and uses the two unaccompanied men to ferry couples across. Similarly to the

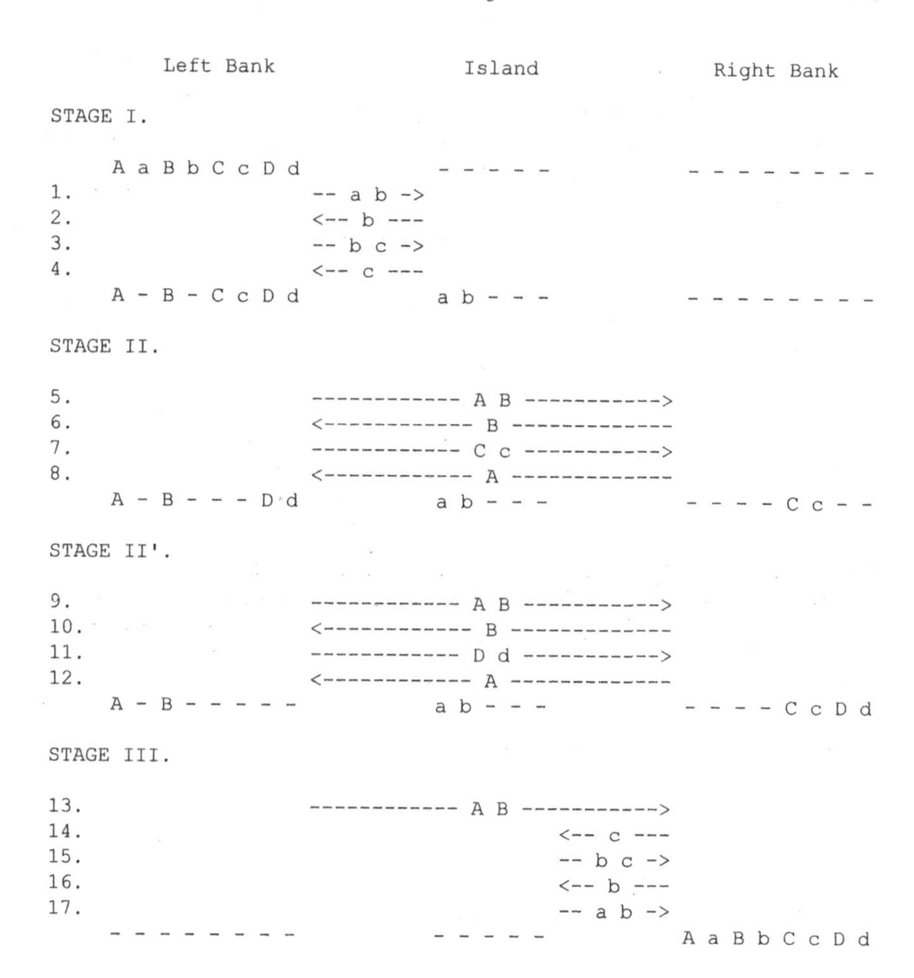

Figure 14.5. The $4n + 1$ solution.

solution of a variant problem where one has two children and some adults and the boat holds the two children or one adult.

Lucas' 1895 solution [14] has this basic technique, but fails to get the beginning and end optimized and gets $4n + 5$ crossings.

In trying to understand the somewhat anomalous value for $n = 4$, we have found a general solution in $5n - 4$ crossings, which gives the optimal solution for $n = 4$ (and also for $n = 3$). Below we give these solutions. In both solutions, there are three stages and the case $n = 4$ shows how the general case can be formed. Observing that Stage II moves a couple across the river, we see that we can repeat it $n - 3$

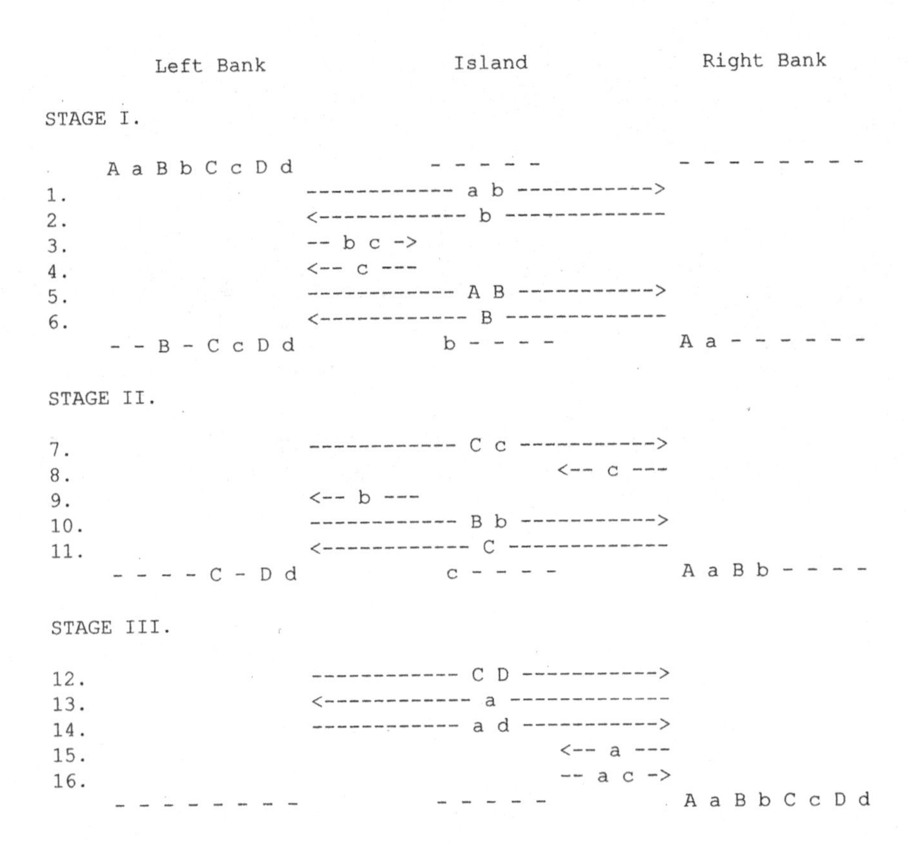

Figure 14.6. The $5n - 4$ solution.

times to move n couples in $5n - 4$ crossings. The ingenious step is the "broken crossing" at moves 8 and 9. This allows us to make one fewer visit to the island (for $n = 4$) than the previous method. (This will be the key step for the solution of the missionaries and cannibals problem in Section 14.5.)

14.4 Proof of Optimality

In this section we prove Theorem 3, below. This requires some notation. Let L, I, R denote the three locations and let N_L, N_I, N_R denote the number of arrivals at and departures from the location. Let N_{LI}, N_{LR}, N_{IR} denote the number of edges or trips (in either direction) between L and I, etc. Then $N_L = N_{LI} + N_{LR}$, $N_I = N_{LI} + N_{IR}$,

$N_R = N_{LR} + N_{IR}$. Now let T be the total number of trips or crossings. Then $2T = 2N_{LI} + 2N_{LR} + 2N_{IR} = N_L + N_I + N_R$. Now $N_L \geq 4n-3$ by the argument given in Theorem 1. Likewise, $N_R \geq 4n-3$, which gives $2T \geq 8n - 6 + N_I$ or

$$T \geq 4n - 3 + N_I/2,$$

where $N_I/2$ is the number of visits to the island.

Since we already have a general solution in $4n+1$ (or $4n$ for $n = 4$) crossings, we now want to show that there is no solution with fewer crossings. If there is any such solution, there is also a minimal one, so we need only look for a minimal solution. Henceforth we shall suppose that we have such a solution and derive consequences that show that it cannot exist.

Bachet pointed out that $N_I/2 = 0$ is impossible for $n = 4$. It is easily verified that this also holds for $n \geq 4$. And $N_I/2 = 1$ is clearly impossible in a minimal solution as a single visit to the island either abandons someone or is a wasted trip. This establishes the following.

Proposition 2. $N_I/2 \geq 2$, *hence* $T \geq 4n - 1$.

So we can now restrict ourselves to minimal solutions with $4n - 1$ or $4n$ (for $n > 4$) crossings and hence with $N_I/2 = 2$ or 3. We will call these interesting solutions.

Consider any visit to the island. This can change the number left on the island by at most one. Hence, if there are ever two people left on the island (i.e. while the boat has gone away from the island), there must be at least four visits to the island, which makes $T \geq 4n + 1$, so this is uninteresting.

Consider again the argument of Theorem 1. Each departure from and return to L can result in the number of people on L being reduced by at most one, except the last departure can reduce the number by 2 to 0. This tells us that $N_L \geq 2(2n - 2) + 1$. If any departure and return fails to diminish the number at L, then there are two extra crossings at L, i.e. $N_L \geq 4n - 1$ and hence $T \geq 4n - 2 + I/2$. This can occur if the departing boat has only one person in it or if the returning boat has two people in it. We shall call such crossings *sub-optimal.* Both sub-optimalities occur if and only if the number at L actually increases and then we have to have another two extra crossings at L, i.e. $N_L \geq 4n + 1$, which gives $T \geq 4n + 1$, so we are not interested in such a solution.

Sub-optimality can also occur at R — if an arrival has only one person or a departure has two people. A bank to bank crossing is sub-optimal at one end if and only if it is sub-optimal at the other end and hence causes two sub-optimalities, which gives $T \geq 4n+1$ and hence is not interesting. However, if we go from one bank to the other via the island, it is possible for this to be sub-optimal at just one bank. If we have one sub-optimality, then we must have $N_I/2 = 2$ for an interesting solution. As at L, the number of people at R after each departure cannot be less than before the arrival without becoming uninteresting. Consider now the situation after four people have left from L and the boat has returned to L. In order that the position at L be legal, at least two women must have gone and if a husband has gone, then his wife must also have gone. Up to equivalence, the departed persons must be: `AaBb, Aabc, abcd`. Since at most one person is on the island, the distribution Island: Right bank can only be one of the following.

(i) `-:AaBb`
(ii) `a:ABb`
(iii) `A:abc`
(iv) `-:abcd`
(v) `a:bcd`

We claim that positions (iii), (iv) and (v) cannot lead to a solution. We can happily continue to move women to R, but we cannot reduce the number of women below three without becoming uninteresting. Then there is no way we can get a man to R. He can only arrive in a boat with one man, two men or a couple. But there will be unescorted women at R, which is not permitted.

Now consider position (i). What could the position on R be before the boat last left? If two people left, then two people also arrived back at L (possibly with one stopping off at I en route and being retrieved later) and so we have a doubly sub-optimal situation which is not interesting. Hence only one person left. If this was a woman, then the position was illegal. If it was a man, then his wife cannot have been left unescorted at L, so she must be at I and she is returned to L as or just after the man returns to L. If the man picks her up, then his return is sub-optimal and the island has already been visited at least twice, so there can be no further visits to the island and no further sub-optimal crossings. Then from position (i), only a couple can cross from L and there is no legal way to return a man to L, except if $n = 4$, when we can send both remaining men from L to R and then ferry the women across. If the man did not pick up his wife from the island, then she was picked up later and thus the number of people at L was increased by this and hence the situation is uninteresting. (In fact, we think that one cannot even get to position (i) without a sub-optimality.)

Finally, consider position (ii). What could the position on R be before the boat last left? If two men or a couple left, then they must have gone to L and this is a double sub-optimality, hence uninteresting. If two women left, then the position at R was illegal. If one man left, then the position at L had been illegal. So only one woman can have left and that cannot be `b, c, d,` ..., so it must have been `a`! This could only happen if `a` leaves R and gets out at the island, where one of `c, d,` ... takes the boat on to L. This means that the island was visited at least once before, is visited once now and must be visited at least once again to retrieve `a`. Hence an interesting solution involving position (ii) can have no suboptimal crossings. Now consider what can happen after position (ii). Any way of moving two women from L leads to an illegal position at R. Any way of moving two men leaves an illegal position at L, except when $n = 4$, which leads to the $5n - 4 = 16$ crossing solution. We now suppose $n \geq 5$. So we would have to move a couple, say `Cc`, to R. If we now move a woman from R, we will get an illegal position at L. Hence we can only move `A` back to R. If he stops at the island to pick up his wife, then their arrival at R is sub-optimal and hence this is uninteresting. So he must go directly to L, giving us now:

```
A------DdEe  -a---------  --BbCc----
```

Neither two men nor two women can now cross, so only a couple can cross, giving:

```
A--------Ee  -a---------  --BbCcDd--
```

Now no men can leave from R and sending a couple or two women back is wasteful, so only a single woman can return and she must stop at the island to let `a` return to L. Hence if $n \geq 5$, we must make at least 4 visits to the island and this is uninteresting. However, this analysis did lead us to the $5n-4$ solution. This completes the proof of our main result.

Theorem 3. *The minimal number of crossings to ferry n couples across a river with an island, using a two person boat and permitting bank to bank crossings, is* $4n+1$ *for* $n \geq 5$ *and is* $5n-4$ *for* $n = 3, 4$.

Looking through Sam Loyd's work [13] we find another solution which I had forgotten. He has 4 couples, a 2 person boat, with an island and the stronger constraint that no man is to get into the boat alone if there is a girl, not his partner, alone on either the island or the other shore. "The [problem] presents so many complications that the best or shortest answer seems to have been overlooked by mathematicians and writers on the subject." "Contrary to published answers, ... the feat can be performed in 17 trips, instead of 24." His solution is the same as that given by Dudeney, but he does not assert that the solution is minimal.

We have checked all the methods used above and find that all of them satisfy Loyd's stronger constraint. But if we make the constraint stronger so that a man or two men cannot get into the boat if there is a female, without her husband and who is not the wife of a man in the boat, on either the island or the other shore, even if escorted by another woman, then our improved solutions might be considered unsafe.

In the $4n+1$ solution, consider crossing 6. Here B goes from bank to bank while a and b are on the island. Could B detour and assault a? — or do we deem the presence of b adequate to defend a's virtue? Similarly, crossings 7, 8, 10, 11, 12 might be considered unsafe. In the $5n-4$ solution, one might similarly consider the crossings 7 and 10 unsafe. On the other hand, both De Fontenay's $8n-6$ solution and Dudeney's $6n-7$ solution are free from these generalized unsafe crossings. Dudeney's solution may not be minimal for these more stringent conditions.

For example, a man is not permitted to row from bank to bank when there are two ladies on the island. Then all the improved solutions in [4] break down, while the longer solution given by Loyd & Dudeney does satisfy the more stringent constraints. We do not know if the Loyd & Dudeney method is minimal for $n > 4$.

Problem 18 has become even better known than problem 17. The problem can be extended to more objects, but then one of the objects must be able to row. A version with four objects appears in Gori's book of 1571 [11, 21]. I found that one could extend this to a father with n children, where children will fight with the children adjacent in age to them unless the father is present and the oldest child can row [20].

14.5 Missionaries and Cannibals

There are several variants of our problem which have appeared relatively recently. Ball [7], from the 3rd ed. of 1896, reports that G. Tarry (early editions had H. Tarry) suggests the problem with an island and n men traveling with their harems of m wives each. Obviously, this would be in some culture where women were not permitted to row! Contemplation indicates that this may not be a serious problem proposal.

However, it seems to be the first example where the ability to row is introduced as a further condition. For the traditional problem with $n = 3$ couples and no island, one needs 3 people who can row in order to get a solution in 11 crossings. If there are only 2 rowers, a solution takes 13 crossings. We have not yet pursued this topic, beyond noting that the $4n + 1$ solution can be done with n rowers: `A, B, c` and one from each other couple.

Pacioli seems to be the first to suggest use of a larger boat. The fifth edition of Bachet discusses this, but Lucas [15] (vol. 1, p. 11) suggests that this material is due to the 19th century editor.[c] Delannoy and/or Lucas [15] (vol. 1, pp. 221–222) suggest the problem of minimizing the number of crossings to ferry n couples (with no island) in an x person boat and they give some results. We can, of course, further complicate the problem by introducing conditions on how many persons can row. We have not investigated these problems.

Of more interest is a change of character of the problem. The first example that we know of is in the 1624 book under the name of van Etten, but popularly attributed to Leurechon [22]. Here there are three masters and three valets, but the men hate the other valets and

[c] I have since seen the second edition which indeed does not mention a larger boat.

will beat them if left with them. This is exactly the problem of the jealous husbands except that it is now based on class discrimination rather than sex discrimination. The masters and valets formulation is also in versions of Ozanam from 1725 [17].

In 1881, Cassell's *Book of In-Door Amusements* gives a reversal of the situation where the servants are dishonest and will rob the masters if the servants ever outnumber the masters at any location. This makes a subtle change in the problem, in that the connection of a master to a particular servant is no longer relevant and we will study this below. By the end of the 19th century, the problem was stated in the form of missionaries and cannibals, where the cannibals must never outnumber the missionaries. The earliest version of this form that we know of to know of is [Pocock] which describes the problem as well known in 1891 while [Sanford] remarks that it is a modern variant in 1927. This formulation has remained popular and has also been varied to explorers and natives (and capital and labor).

We have since found a version of the problem, with kings and dishonest servants, in [16]. An earlier version of the missionaries and cannibals version in the context of masters and servants occurs in: [12] (Arithmetical Puzzles No. 21, pp. 5 & 56). Here the servants will murder their masters if they ever outnumber them.

The numerical conditions discussed in our Section 4 are a complete description of the permissible situations in this version, as in the original version, but there is a subtle difference in the permissible moves which we can see by looking at moves 8 and 9 in the $5n - 4$ solution. Before these moves, we have:

```
– – B – – – Dd    – – – b – – – –    Aa – –Cc – –
```

and after, we have:

```
– – Bb – –Dd    – – – – – c – –    Aa – –C – ––.
```

If we have missionaries and cannibals represented by `M`, `C`, then the positions are:

```
MMC – – – – –    – – – – C – – –    MMCC – – – –
```

and:

```
MMCC – – – –    – – – –C – – –    MMC – – – ––.
```

In the latter version, we can accomplish the transition by a single crossing of a cannibal from R to L, but it takes two crossings in the original version. Any solution of the jealous husbands problem gives a solution of the missionaries and cannibals problem, but this observation shows that the converse need not hold. In other words, the missionaries and cannibals problem is a proper weakening of the jealous husbands problem.

Indeed, our observation has shown that the missionaries and cannibals problem with $n = 4$ has a solution in 15 crossings. Further, since this observation reduces Stage II of the $5n-4$ solution to having only 4 crossings, it follows that this gives us a solution in $4n-1$ crossings for $n \geq 4$. It is easily verified that an island is indeed necessary for $n \geq 4$ and so the argument leading to Proposition 2 shows that $4n - 1$ is minimal. This also gives the correct number when $n = 3$, though then no island is needed. Clearly the solution with no bank to bank crossings cannot be reduced as it already uses the minimum possible number of crossings, namely $8n - 6$. We state these results formally.

Theorem 4. *First, the minimal number of crossings to ferry n missionaries and n cannibals across a river with an island, using a two person boat and with no bank to bank crossings, is $8n - 6$ moves for $n \geq 1$. Second, the minimal number of crossings to ferry n missionaries and n cannibals across a river with an island, using a two person boat and permitting bank to bank crossings, is $4n - 1$ moves for $n \geq 3$.*

(The observations surrounding Loyd's variation do not affect the missionaries and cannibals version of the problem. In translating a solution to the latter version, the missionaries replace the husbands and the cannibals replace the women and it is unlikely that a single missionary in the boat will detour to the island to provide a feast for the two cannibals parked there!)

14.6 Other Cultures

Some recent work on the problem in other cultures is contained in the following. Zaslavsky [23] says that leopard, goat and pile of cassava leaves is popular with the Kpelle children of Liberia. However, this is based on an ambiguous description and an earlier report of a Kpelle version has the form described below [5] (p. 120).

Ascher [4] describes many appearances in folklore of many cultures. She discusses African variants of the wolf, goat and cabbage problem in which the man can take two of the items in the boat. This is much easier, requiring only three crossings, but some versions say that the man cannot control the items in the boat, so he cannot have the wolf and goat or the goat and cabbage in the boat with him. This still only takes three crossings. Various forms of these problems are mentioned: fox, fowl and corn; tiger, sheep and reeds; jackal, goat and hay; caged cheetah, fowl and rice; leopard, goat and leaves. She also discusses an Ila (Zambia) version with leopard, goat, rat and corn which is unsolvable!

Ascher's *Ethnomathematics* [5] is a good survey of the problem and numerous references to the folklore and ethnographic literature and amplifies the above article. A version like the wolf, goat and cabbage is found in the Cape Verde Islands, in Cameroon and in Ethiopia. The African version is found as far apart as Algeria and Zanzibar, but with some variations. An Algerian version with jackal, goat and hay allows one to carry any two in the boat, but an inefficient solution is presented first. A Kpelle (Liberia) version with cheetah, fowl and rice adds that the man cannot keep control while rowing so he cannot take the fowl with either the cheetah or the rice in the boat. A Zanzibar version with leopard, goat and leaves adds instead that no two items can be left on either bank together. (A similar version occurs among African-Americans on the Sea Islands of South Carolina.)

Bibliography

[1] AHRENS. Vol. 1, 1–13; Vol. 2, 315–318.
[2] ABBOT ALBERT. 333–334.
[3] ALCUIN. 54–56.
[4] M. Ascher. "A river-crossing problem in cultural perspective." *Mathematics Magazine*, 63 (1990) 26–28.
[5] M. Ascher. *Ethnomathematics.* Brooks/Cole Publishing, 1991. Section 4.8, pp. 109–116 & Note 8, 119–121.
[6] BACHET.
[7] BALL.
[8] *Brandreth Puzzle Book.* Brandreth's Pills, 1896.

[9] COLUMBIA ALGORISM. The plates between 402 and 403 which give the illustrations of both crossing problems.
[10] H. E. Dudeney. "Prob. 376." In *Amusements in Mathematics*, Nelson, 1917, 113 & 237. (also Dover, 1958.)
[11] DIONIGI GORI. Libro di arimetricha. 1571. Italian MS in Biblioteca Comunale di Siena, L. IV. 23. Extensively quoted and discussed in: R. Franci & L. Toti Rigatelli; Introduzione all'Aritmetica Mercantile del Medioevo e del Rinascimento; Istituto di Matematica dell'Università di Siena, nd [1980?]. (Later published by Quattroventi, Urbino, 1981.) (I will quote Gori's folios and also give the pages of this Introduzione.)
[12] JACKSON.
[13] SAM LOYD. *Sam Loyd's Cyclopedia of 5,000 Puzzles, Tricks and Conundrums* (ed. by Sam Loyd II). (Lamb Publishing, 1914) = Pinnacle or Corwin, 1976. The four elopements, pp. 266 & 375. [This book is a reprint of Loyd's *Our Puzzle Magazine*, a quarterly which started in June 1907. This material would have appeared in 1:4 (Apr 1908).]
[14] LUCAS. "Les vilains maris jaloux," 125–144 & "Note II—Sur les traversées," 198–202.
[15] LUCAS. Vol. 1, 1891, 1–18 & 221–222; vol. 2, 1883, 240.
[16] L. Mittenzwey. *Mathematische Kurzweil.* Julius Klinkhardt, 1880. There were several reprintings. In 1880, it is prob. 228, 42 & 92. In the 4th ed., 1904, it is prob. 255, 46 & 94. In the 7th ed., 1918, it is prob. 255, 42 & 90.
[17] JACQUES OZANAM. *Recreations Mathematiques et Physiques.* Nouv. ed., 4 vols., C. A. Jombert, Paris, 1725, vol. 1, pp. 4–5. {I [DS] have examined more editions of Ozanam. I can't find it in the original 1694 ed. It is Prob. 3, pp. 4–5 in the 1725 ed.; Prob. 19, pp. 171–172 in the 1790 ed., pp. 171–172 in Hutton's 1803 English ed., pp. 150–151 in the 1814 reprint of Hutton; Prob. 18, p. 77 in Riddle's 1840 ed. of Hutton.}
[18] PACIOLI-DVQ.
[19] VERA SANFORD. *The History and Significance of Certain Standard Problems in Algebra.* Teachers College, Columbia Univ., NY, *Contributions to Education*, No. 251, 1927. (Reprinted by AMS Press, NY, 1972.)
[20] D. Singmaster. "Keeping the Peace." *Weekend Telegraph* (7 & 14 July 1990) xxiv & xxii.
[21] TARTAGLIA. Part 1, Book 16, art. 141–143, ff. 257r–257v.
[22] ETTEN (1624: Prob. 14 & 15, pp. 1415.)
[23] C. Zaslavsky. *Africa Counts.* Prindle, Weber & Schmidt, 1973, 109–110.

Chapter 15

Sharing Barrels

> A father, when dying, gave to his sons 30 glass flasks, of which 10 were full of oil, 10 were half-full, and the last 10 were empty. Divide the oil and the flasks, so that each of the three sons receive equally of both glass and oil.

The earliest examples of barrel sharing problems appear to be in the 9th century collection attributed to Alcuin [3]. His problem 12, above, is the standard problem with 10 barrels of each type. Problem 51 is a variant — there are four barrels containing 10, 20, 30, 40 measures of wine and they are to be equally divided among four sons. Some shifting of contents is required. In the 13th century, Abbot Albert [1] gives the problem of dividing nine barrels containing $1, 2, \ldots, 9$ measures among three persons. This is generalized in the final section.

In Bachet [5], we find examples where there are different numbers of barrels of the three types and an example where the barrels must be divided among four persons. Ahrens [2] says that some of this material was added by the 19th century editor — earlier editions of [5] need to be consulted to verify this.

In this chapter, we consider the general problem with N of each type of barrel. The number of solutions will be seen to be the same as the number of triangles with integer sides and perimeter N. These were studied in [4, 11] by use of intricate summations. Their work is explained and extended in [10]. Here a geometric approach is given using triangular coordinates which is easier to understand and brings out several further properties, including the connection between the number of incongruent triangles and partitions into at most three parts. The mathematics in this chapter is not advanced but it is more complex than most of the topics in this book.

15.1 The Barrels Problem

Suppose we have N barrels of each type: full, half-full and empty. Let f_i, h_i, e_i be the number of these which the ith person receives, $i = 1, 2, 3$. These are clearly non-negative integers and we shall assume this from now on. Then we have a fair sharing if and only if the following conditions hold.

$$\begin{aligned} f_i + h_i + e_i &= N, \quad \text{for } i = 1,2,3. \\ f_i + h_i/2 &= N/2, \quad \text{for } i = 1,2,3, \\ \sum_i f_i = \sum_i h_i &= \sum_i e_i = N. \end{aligned} \tag{15.1}$$

A little observation and manipulation shows that (1) implies $e_i = f_i$ and $h_i = N - 2f_i$, and hence that they are solved by knowing the f_i subject to:

$$\begin{aligned} \sum_i f_i &= N; \\ f_i \le N/2, &\quad \text{for } i = 1,2,3. \end{aligned} \tag{15.2}$$

15.2 Integral Triangles

It is well known and easily seen that three non-negative lengths x, y, z can form a triangle if and only if the three triangle inequalities hold:

$$x + y \ge z, \quad y + z \ge x, \quad z + x \ge y. \tag{15.3}$$

If we set $x + y + z = p$, then (3) is equivalent to:

$$x \le \frac{p}{2}, \quad y \le \frac{p}{2}, \quad z \le \frac{p}{2}. \tag{15.4}$$

(The triangle is non-degenerate if and only if the inequalities are all strict.) Hence the solutions for sharing n barrels of each type are just the integral lengths which form a triangle of perimeter n.

15.3 Triangular Coordinates

Consider a triangle of sides x, y, z and perimeter p. Since $x+y+z=p$, we can view (x, y, z) as a point on the plane $x+y+z=p$, in the triangle cut off by the planes $x=0, y=0, z=0$. This gives the standard representation of (x, y, z) in triangular coordinates as shown in Figure 15.1. (Ignore the broken lines in Figure 15.1b for the moment.) Letting the spacing between lines be our unit of distance, the point (x, y, z) is located x units from the right edge, y units from

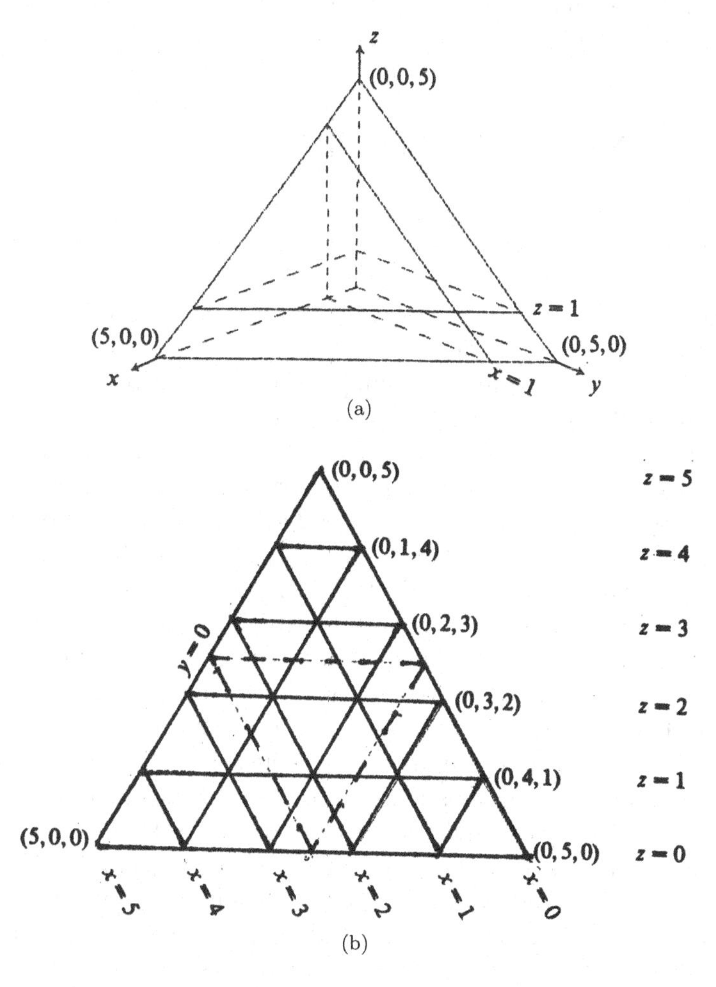

Figure 15.1. Triangular coordinate system.

the left edge and z units from the bottom edge of the triangle. It is a classic property of the equilateral triangle that $x+y+z$ is a constant, namely the altitude. If we consider integral values of x, y, z with an integral sum p, we see that these points (x, y, z) form a triangular array having $p+1$ points along an edge. We denote such an array as $\mathbf{T}(p+1)$. $\mathbf{T}(p+1)$ clearly has $1+2+\cdots+(p+1) = \frac{1}{2}(p+1)(p+2) = T(p+1)$ points, where $T(p)$ denotes the pth triangular number.

The points along the edges of $\mathbf{T}(p+1)$ correspond to at least one of x, y, z being 0, so the interior points correspond to all lengths being positive. These thus form a $\mathbf{T}(p-2)$ with $T(p-2)$ points. Readers will find it useful to draw diagrams as they read on.

15.4 The Number of Integral Triangles

In our triangular coordinates, we see that (x, y, z) corresponds to an integral triangle of perimeter p if and only if it is an integer point in $\mathbf{T}(p+1)$ which lies inside the central region cut off by the conditions: $x \leq \frac{1}{2}p$, $y \leq \frac{1}{2}p$, $z \leq \frac{1}{2}p$, as indicated by the broken lines in Figure 15.1(b) for $p = 5$.

Let $T_1(p)$ be the number of integral triangles of perimeter p and let $T_2(p)$ be the number which are non-degenerate.

If p is odd (as in Figure 15.1(b)), let $p = 2q+1$. Then the central region cut off by our conditions gives a triangle with base on the line $z = q$. This line contains $p+1-q = q+2$ points, but the cut off region omits the two end points, so our region is a $\mathbf{T}(q)$, which contains $T(q)$ points. (Alternatively, take $T(2q+2) - 3T(q+1)$ to obtain $T(q)$.) All of these points correspond to non-degenerate triangles, so we have shown that $T_1(p) = T_2(p) = T(q)$. These are precisely the solutions of our barrel sharing problem for p barrels of each type. If p is even, let $p = 2q$. Then the central region cut off our conditions is a triangle whose base is the whole line $z = q$, hence is a $\mathbf{T}(q+1)$ and we have $T_1(p) = T(q+1)$. This is the number of solutions of our barrel sharing problem, since we do not restrict ourselves to non-degenerate solutions. But our central region certainly does contain degenerate triangles. We can remove all of these by excluding the lines $z = q$, $y = q$, $x = q$. This leaves a central region which is a $\mathbf{T}(q-2)$, so $T_2(p) = T(q-2)$. (As before, these results can be obtained by subtracting from $T(p+1)$.)

Note that both $p = 2q - 2$ and $p = 2q + 1$ give the same central region $\mathbf{T}(q)$ of integer points corresponding to triangles of perimeter p, while both $p = 2q + 1$ and $p = 2q + 4$ give the same central region $\mathbf{T}(q)$ of integer points corresponding to non-degenerate triangles of perimeter p. The latter half of the last sentence is the geometric basis of Theorem 3 in [11]. From these observations, we see that $T_1(p) = T_2(p + 3)$. This is also easily seen since adding one to each length gives a one-to-one correspondence between the triangles being counted.

15.5 The Number of Incongruent Integral Triangles

In enumerating the solutions of the barrel sharing problems, we do not really care which person gets which share, since each share is fair. If x, y, z is a fair distribution of full barrels, then we consider this as equivalent to y, x, z, etc. (all six permutations of x, y, z are considered as equivalent solutions).

Viewing x, y, z as sides of a triangle, there are six ways in which it can be congruent to another triangle. That is, one triangle is congruent to another if and only if the sides of one are a permutation of the sides of the other. These correspond to the six permutations of x, y, z and to the six symmetries of our triangular region.

So to count the number of inequivalent solutions of the barrel sharing problem or to count the number of incongruent integral triangles, we need to count the points of our central triangular region which are inequivalent under the symmetries of the triangle. Let $T_3(p)$ be the number of incongruent integral triangles of perimeter p and let $T_4(p)$ be the number which are non-degenerate. Let $N(q)$ be the number of inequivalent points in $\mathbf{T}(q)$. Then $T_3(2q - 2) = T_3(2q + 1) = N(q)$ and $T_4(2q+1) = T_4(2q+4) = N(q)$. Again, there is a shift of three between the general case and the non-degenerate case, i.e. $T_3(p) = T_4(p + 3)$.

Theorem 1. $N(q + 3) = N(q) + \lfloor \frac{q+4}{2} \rfloor$.

Proof. The array $\mathbf{T}(q+3)$ is obtained by bordering $\mathbf{T}(q)$. The new inequivalent points are those in the border and they comprise half of a bordering edge. Such an edge has $q + 3$ points and we must count

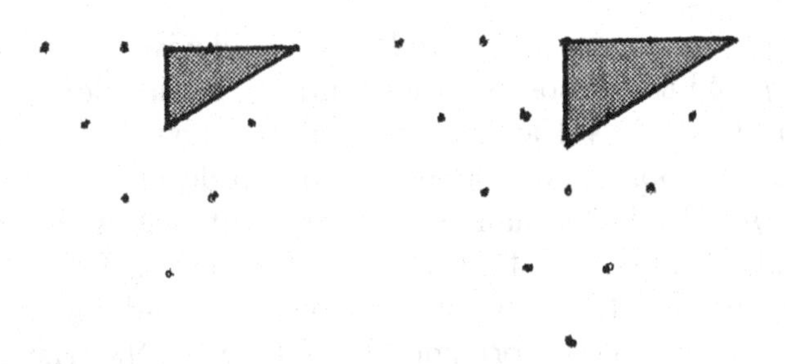

Figure 15.2. Inequivalent points for $q + 3 = 4, 5$.

the midpoint when $q+3$ is odd, giving $\lfloor \frac{q+4}{2} \rfloor$ new inequivalent points, Figure 15.2. □

Corollary 1.1. *The sequence $(N(q))$ is determined by the recurrence in Theorem 1 and the initial conditions: $N(1) = N(2) = 1$, $N(3) = 2$. These values can be extended backward, consistent with the theorem, to $N(0) = N(-1) = N(-2) = N(-3) = N(-4) = 0$.*

Corollary 1.2. $N(q + 6) = N(q) + q + 5$.

Repeated use of Corollary 1.2 gives us the following.

Corollary 1.3. *Let $q - 1 = 6k + r$, with $0 \leq r < 6$.*
If $r = 0$, then $N(q) = 6T(k) + 1 = 3k(k + 1) + 1$.
If $r \neq 0$, then $N(q) = 6T(k) + r(k + 1) = (3k + r)(k + 1)$.

This corollary holds for $q \geq -4$ and can be extended backward. Corollaries 1 and 3 contain Theorems 1 and 2 of [11], but seem much simpler to me. See Table 15.1.

15.6 Relation to Partitions

Looking up the sequence $N(q)$ in Sloane's invaluable handbook [14], one finds that it is the same as the partitions of $q - 1$ into at most three parts. To see this, view $\mathbf{T}(q)$ as the points (x, y, z) such that $x+y+z = q-1$. Then taking just the inequivalent points is precisely the same as taking the partitions of $q-1$ into at most three parts. Let $P_d(n)$ denote the number of partitions of n into at most d parts, so we have $N(n + 1) = P_3(n)$. Then Theorem 1 is a form of the known

Table 15.1. Recall $p = 2q$ or $2q + 1$.

p	q	$T(p)$	$N(p)$	$T_1(p)$	$T_2(p)$	$T_3(p)$	$T_4(p)$
0	0	0	0	1	0	1	0
1	0	1	1	0	0	0	0
2	1	3	1	3	0	1	0
3	1	6	2	1	1	1	1
4	2	10	3	6	0	2	0
5	2	15	4	3	3	1	1
6	3	21	5	10	1	3	1
7	3	28	7	6	6	2	2
8	4	36	8	15	3	4	1
9	4	45	10	10	10	3	3
10	5	55	12	21	6	5	2
11	5	66	14	15	15	4	4
12	6	78	16	28	10	7	3
13	6	91	19	21	21	5	5
14	7	105	21	36	15	8	4
15	7	120	24	28	28	7	7
16	8	136	27	45	21	10	5
17	8	153	30	36	36	8	8
18	9	171	33	55	28	12	7
19	9	190	37	45	45	10	10
20	10	210	40	66	36	14	8

result that $P_3(n+3) = P_3(n) + P_2(n+3)$. This says that a partition of $n+3$ either has 3 positive parts, and hence arises from a partition counted by $P_3(n)$ by adding 1 to each part, or has a zero part, and hence arises from a partition counted by $P_2(n+3)$ by adding an extra part of 0. We see also that the number of partitions of $n+3$ into exactly three parts (i.e. with no zero parts) is just $P_3(n)$.

We have seen that $T_3(2n-2) = N(n)$ and that the latter is $P_3(n-1)$. We can see this another way: $T_3(2n-2)$ counts those triples x_1, x_2, x_3 such that $x_1 + x_2 + x_3 = 2n-2$, with $0 \leq x_i \leq n-1$. Letting $y_i = n - 1 - x_i$, we have that $y_1 + y_2 + y_3 = n - 1$, with $0 \leq y_i \leq n-1$. Hence the triple y_1, y_2, y_3 is a partition of $n-1$.[a]

[a]The pretty correspondence between x_i and y_i has occurred to several people. It is in my unpublished 1982 paper on integral triangles and was also found by both N. J. Fine and P. Pacitti [10, pp. 45–46].

In the context of barrel sharing, when $N(n) = 2n-2$, then the x_i are the f_i of Section 15.1 and so $y_i = \frac{1}{2}n - f_i = \frac{1}{2}h_i$. This shows that, for even n, the sharing of barrels is determined by sharing the $\frac{1}{2}n$ pairs of half-full barrels in any way. Similar arguments apply for the odd case and for the non-degenerate cases. For sharing $N = 2n+1$ barrels of each type, each person must receive an odd number of half-full barrels. Thus the sharing is determined by giving each person one half-full barrel and then distributing the remaining $\frac{N-3}{2} = n-1$ pairs of half-full barrels in any way. Thus $T_3(2n+1) = T_3(2n-2) = P_3(n-1)$.

In [4] (and [10]), it is shown that $P_3(n-3) = [\frac{n^2}{12}]$, where $[x]$ is the nearest integer to x, and hence that $T_4(n) = [\frac{n^2}{12}] - \lfloor\frac{n}{4}\rfloor\lfloor\frac{n+2}{4}\rfloor$. The reader can ponder the connection between this and our results: $T_4(2q+1) = T_4(2q+4) = N(q) = P_3(q-1)$, Theorem 1 and its corollaries.

15.7 Other Versions

If we have F full barrels, H half-full barrels and E empty barrels, then conditions become the following.

$$f_i + h_i + e_i = \frac{F+H+E}{3}, \quad \text{for } i = 1,2,3,$$

$$f_i + \frac{h_i}{2} = \frac{F+\frac{1}{2}H}{3}, \quad \text{for } i = 1,2,3,$$

$$\sum_i f_i = F, \quad \sum_i h_i = H, \quad \sum_i e_i = E. \tag{15.5}$$

When is there an integral solution? The existence of an integral solution imposes certain constraints on F, H, E, namely that $2F+H$ and $F+H+E$ must be divisible by 3. These are easily seen to be equivalent to: $F \equiv H \equiv E \pmod 3$. However, we already know that $F = H = E = 1$ has no solution, but looking closer gives the following.

Theorem 2. *There is a fair sharing of F full, H half-full and E empty barrels among three people if and only if $F \equiv H \equiv E \pmod 3$ and $H \neq 1$.*

We do not prove this as it is a special case of Theorem 3 below.

One might think that the number of solutions of the above equations could be found since any solution would be given by knowing the f_i subject to:

$$\sum_i f_i = F,$$

$$f_i \le \frac{F + \frac{1}{2}H}{3}, \quad \text{for } i = 1, 2, 3. \tag{15.6}$$

However, one must also have $0 \le f_i \le F$ and $f_i \le \frac{F+H+E}{3}$ and further, that $0 \le h_i \le H$, $h_i \le \frac{2F+H}{3}$, $h_i \le \frac{F+H+E}{3}$ and $0 \le e_i \le E$, $e_i \le \frac{F+H+E}{3}$. These 11 sets of inequalities (counting also $0 \le f_i$) give a rather complex set of conditions on the f_i and the same holds if we try to express solutions in terms of the h_i or e_i.

If $F = H = E = N$ and we share among k people:

$$f_i + h_i + e_i = \frac{3N}{k}, \quad \text{for } i = 1, 2, \ldots, k,$$

$$2f_i + h_i = \frac{3N}{k}, \quad \text{for } i = 1, 2, \ldots, k,$$

$$\sum_i f_i = \sum_i h_i = \sum_i e_i = N. \tag{15.7}$$

Again, a solution is determined by knowing the f_i, now subject to simple conditions:

$$\sum_i f_i = N;$$

$$f_i \le \frac{3N}{2k}, \quad \text{for } i = 1, 2, \ldots, k. \tag{15.8}$$

Geometrically, this leads to simplicial coordinates in $k-1$ dimensions, but the problem is no longer the same as finding k integral lengths which form a k-gon of perimeter N, for which the conditions are:

$$\sum_i f_i = N;$$

$$f_i \le \frac{N}{2}, \quad \text{for } i = 1, 2, \ldots, k. \tag{15.9}$$

It is possible to generalize and extend the previous ideas to find the number of inequivalent solutions of these last equations, but it is not very illuminating and does not give the simple connection with partitions that occurs for $k = 3$. Further, this is not the number of incongruent integral k-gons of perimeter N, since, e.g. this considers a, b, c, d as the same as b, a, c, d and since a quadrilateral with sides a, b, c, d has infinitely many incongruent shapes.

Obviously, one can combine both of Bachet's ideas and try to divide F, H, E among four or k persons. Ozanam [13] gives a confused version of this — he seems to start with $F = H = E = 8$, divided among four people, but gives a solution for $F = E = 6, H = 12$, though he seems to distinguish 6 half-full barrels from 6 half-empty barrels. Some trial and error leads to the following.

Theorem 3. *There is a fair sharing of F full, H half-full and E empty barrels among k people if and only if: (a) $F \equiv E \pmod{k}$; (b) $H \equiv -2F \pmod{k}$; (c) if $\frac{2F+H}{k}$ is odd, then $H \geq k$.*

Proof. The conditions for a fair sharing are: (d) $f_i + h_i + e_i = \frac{F+H+E}{k}$, for $i = 1, 2, \ldots, k$; (e) $2f_i + h_i = \frac{2F+H}{k}$, for $i = 1, 2, \ldots, k$; (f) $\sum_i f_i = F, \sum_i h_i = H, \sum ie_i = E$.

From (d) and (e), we get $f_i - e_i = \frac{F-E}{k}$ for each i, so that (a) and (b) must hold if there is a solution. If $\frac{2F+H}{k}$ is odd, then (e) shows that h_i is odd, hence $h_i \geq 1$, for each i. Hence $H \geq k$ and the "only if" part of the theorem is proven.

Suppose that conditions of the theorem hold. Let $F \equiv f \pmod{k}$, with $0 \leq f < k$. If $f = 0$, then we have $F \equiv H \equiv E \equiv 0 \pmod{k}$ and there is an easy solution. Suppose now that $f > 0$. Distribute 1, 0, 1 (i.e. 1 full, 0 half-full and 1 empty barrels) to f people and 0, 2, 0 to the remaining $k - f$ people. This leaves $F - f, H - 2(k - f), E - f$ barrels. We have $F - f \equiv E - f \equiv 0$ and $H - 2(k - f) \equiv H + 2f \equiv 0 \pmod{k}$, so these remaining barrels can be easily shared. So we will have a fair sharing, provided only that $H \geq 2(k - f)$, which we rewrite as $\frac{2f+H}{k} \geq 2$. Since $f > 0$, we have that $\frac{2f+H}{k} > 0$. If $\frac{2f+H}{k} = 1$, then also $\frac{2f+H}{k}$ is odd and condition (c) says that $H \geq k$, which gives $\frac{2f+H}{k} > 1$. Hence $\frac{2f+H}{k} \geq 2$ and our distribution can indeed be carried out to give a fair sharing. □

Note that for $k = 3$, we have $-2 \equiv 1 \pmod{k}$, so that conditions simplify to give the conditions in Theorem 2.

Kraitchik [12] has varied the problem still further by having 9 barrels of each of the following five types: full, $\frac{3}{4}$ full, $\frac{1}{2}$ full, $\frac{1}{4}$ full and empty, to be divided among 5 people!

15.8 Fair Division of the First kn Integers into k Parts

In the 13th century *Annales Stadenses*, compiled by Abbot Albert [1], there is an insertion of thirteen recreational problems. In Chapter 5, there is an annotated translation of these. The third problem, repeated here, is both novel and forgotten. It will be generalized after giving the history.

> Firri said further: There were three brothers at Cologne, who had nine casks of wine. The first cask contained 1 bucket, the second 2, the third 3, the fourth 4, the fifth 5, the sixth 6, the seventh 7, the eighth 8, the ninth 9. Divide this wine equally among these three, without breaking any casks. I shall do it, said Tirri. To the oldest, I give the first, fifth and ninth, and he has 15 buckets. To the middle one, I give the third, fourth and eighth, and he likewise has 15. So to the youngest I give the second, sixth and seventh; and thus he also has 15, the wine is divided and the casks are not broken. [Though not clearly specified, the problem wants the casks to be equally divided as well as the wine.]

Clearly this is a special case of the following problem.

> For integers k, n, can one partition the first kn integers into k sets of size n such that each set has the same sum?

We call this a fair division of the first kn integers into k parts. Albert is the first known appearance of this uncommon variant of the classic problem of sharing barrels.

The classic problem has already been discussed in this chapter. However, Alcuin also has a problem, no. 51, with four barrels containing 10, 20, 30, 40 measures of wine and he divides them among four sons, which requires some pouring of wine. This is clearly a forerunner of Albert's problem, or perhaps Alcuin had misunderstood or miscopied the problem. This may be the earliest version of this problem. After Albert, the identical problem occurs in two well-known manuscripts of the 14th and 15th centuries from Germany [8, 15].

Folkerts [6] found 10 occurrences in anonymous Latin manuscripts from the 13–15 centuries, but only [15] seems to have been published and the problem did not get into later problem collections and essentially disappeared from common knowledge, except for a late 19th century discussion of Albert's problem [9].

Proof of Existence

The sum of the first kn integers is the knth triangle number $T_{kn} = \frac{1}{2}kn(kn+1)$. For a fair division to exist, we must have k divides T_{kn}. It is easily seen that $\frac{1}{k}T_{kn} = \frac{1}{2}n(kn+1)$ is integral if and only if n is even or k is odd, so this is a necessary condition for a fair division to exist. However, the case $n = 1$ is trivially impossible.

Proposition. *If n is even, there is a fair division.*

Proof. The first kn integers can be paired into $\frac{1}{2}kn$ pairs $i, kn+1-i$ with sum $kn+1$. Any way of partitioning these pairs into k sets of $\frac{1}{2}$ pairs solves our problem. □

Example. For $k = 3$, $n = 4$, our pairs are $1, 12$; $2, 11$; $\ldots$; $6, 7$ and a solution is: $1, 12, 2, 11$; $3, 10, 4, 9$; $5, 8, 6, 7$.

Theorem. *If n is odd, $n > 1$ and k is odd, then there is a fair division.*

Proof. We proceed by induction on odd values of n, beginning with $n = 3$. Let $k = 2r+1$, so $\frac{1}{k}T_{3k} = 3(3r+2) = 9r+6$. A solution is given by the following triples.

$$\begin{array}{lll}
1 & 3r+2 & 6r+3 \\
2 & 3r+3 & 6r+1 \\
3 & 3r+4 & 6r-1 \\
\vdots & & \\
r+1 & 4r+2 & 4r+3 \\
r+2 & 2r+2 & 6r+2 \\
r+3 & 2r+3 & 6r \\
r+4 & 2r+4 & 6r-2 \\
\vdots & & \\
2r+1 & 3r+1 & 4r+4
\end{array}$$

Example. For $k = 7$, we have $r = 3$ and a solution: $1, 11, 21$; $2, 12, 19$; $3, 13, 17$; $4, 14, 15$; $5, 8, 20$; $6, 9, 18$; $7, 10, 16$.

The inductive process is clearly seen in going from $n = 3$ to $n = 5$ as follows. The solution above has sums $9r + 6$. We now want quintuples with sums $\frac{1}{k}T_{5k} = 25r + 15$. We add $2r + 1$ to all of the numbers in our previous solution. This gives us triples with sums $9r + 6 + 6r + 3 = 15r + 9$ and these triples use all the numbers from $2r+2 = k+1$ to $8r+4 = 4k$. So it remains to use the first and the last k numbers. But these can be paired $1, 5k$; $2, 5k - 1$; $\ldots$; $k, 4k + 1$; where each pair adds to $5k + 1 = 10r + 6$. Adding these pairs to our augmented triples gives the desired total of $25r + 15$.

Example. For $k = 7$, we can extend from $n = 3$ to $n = 5$ to get a solution: $1, 8, 18, 28, 35$; $2, 9, 19, 26, 34$; $3, 10, 20, 24, 33$; $4, 11, 21, 22, 32$; $5, 12, 15, 27, 31$; $6, 13, 16, 25, 30$; $7, 14, 17, 23, 29$.

Hence we have shown that there is a fair division of the first kn integers into k parts if and only if n is even or (k is odd and $n > 1$). □

Now that we know when a fair division exists, we could ask for the number of them. It is clear from the first few cases that there are lots of fair divisions. Perhaps someone would like to examine this question or to compute a table of the numbers.

Bibliography

[1] ABBOT ALBERT.

[2] W. Ahrens. Altes und Neues aus der Unterhaltungsmathematik. Springer, 1918, p. 29.

[3] ALCUIN.

[4] G. E. Andrews. "A note on partitions and triangles with integer sides." *American Mathematical Monthly* 86 (1979) 477–478.[b]

[5] BACHET. Additional problem 9, pp. 168–171.

[6] M. Folkerts. "Mathematische Aufgabensammlungen aus dem ausgehenden Mittelalter Ein Beitrag zur Klostermathematik des 14. und 15. Jahrhunderts." *Sudhoffs Archiv,* 55 (1971) 58–75. See Section 9 (b) on p. 72.

[b] This is based on [11].

[7] M. Folkerts. "Zur Frühgeschichte der magischen Quadrate in Westeuropa." *Sudhoffs Archiv.* 65 (1981) 313–338. See 315, note 11.

[8] GERHARDT. Prob. 351, 54, with Vogel's notes on page 182. Identical to Abbot Albert, but with the solution arranged in columns. Vogel comments that this makes a "half-magic square" as noted by Günther.

[9] S. Günther. "Geschichte des mathematischen Unterrichts im deutschen Mittelälter bis zum Jahre 1525. " *Monumenta Germaniae Paedagogica III.* 1887. (also Sändig Reprint Verlag, 1969. 35–36.)[c]

[10] R. Honsberger. *Mathematical Gems III.* Math. Assoc. Amer., 1985, 39–47.

[11] J. H. Jordan, R. Walch and R. J. Wisner. "Triangles with integer sides." *American Mathematical Monthly* 86 (1979) 686–689.

[12] M. Kraitchik. *Mathematical Recreations.* Allen & Unwin, 1943. Chap. 2, prob. 34, 31–32. (The second edition has few changes and has been reprinted by Dover, 1953.)

[13] OZANAM-GRANDIN. Vol. 1, prob. 44, 242–246.

[14] N. J. A. Sloane. *A Handbook of Integer Sequences.* Academic Press, 1973, 46: sequence 186.[d]

[15] MUNICH. Prob. XXIV, f. 32r. Same as Abbot Albert and with the same solution.

[16] D. Singmaster. "Triangles with integer sides and sharing barrels." *College Math. J.* 21, 4 (September 1990) 278–285.

[c]He discusses Abbot Albert and this problem, noting that the solution can be viewed as a set of lines in a magic square so that the perpendicular lines give a second solution, but that magic squares were then unknown in Europe. He gives no other examples.

[d]There is now an online version, called OEIS.

Chapter 16

Vanishing Area Paradoxes

In Memoriam Martin Gardner

One of Sam Loyd's most famous puzzles is "The Vanishing Chinaman" or "Get off the Earth." Martin Gardner discussed these extensively in [6].[a]

However, the term "vanishing area puzzle" is used for two different types of puzzle. In "The Vanishing Chinaman" no actual area vanishes — it is one of the figures in the picture that vanishes, so perhaps we should call this a "vanishing object puzzle". In a true "vanishing area puzzle", an area is cut into several pieces and reassembled to make an area which appears to be larger or smaller than the original area. These are considerably older than vanishing object puzzles. Martin was fond of these puzzles — indeed, his first puzzle book [6] devotes two chapters to such puzzles — possibly still the best general survey of them — and he also wrote several columns about them [7, 8]. The most famous version of these is the "Checkerboard Paradox", see Figure 16.2, where an 8×8 checkerboard is cut into four parts and reassembled into a 5×13 rectangle, with a net gain of one unit of area. This chapter is primarily concerned with the early history of such puzzles.

[a]This is based on a talk for International Puzzle Party 20, Los Angeles, 2000. A version appeared in *Recreational Mathematics Magazine* (Lisbon) electronic version 1 (March 2014) 10–21.

Figure 16.1. Sam Loyd Example.

The oldest known version of this is an actual puzzle, dated c. 1900, [1], shown in [18] and there is a 1901 publication [5]. Recently Dario Uri has researched French puzzles of this era and finds that AWGL was Alphonse Wogue and G. Levy, who were active in 1881–1903.

Loyd asserts he presented "this paradoxical problem" to the First American Chess Congress in 1858, but it is not clear if he means the area 65 version or the area 63 version. Loyd would have been 17 at the time. If this is true, he is ten years before any other appearance of the area 65 puzzle and about 42 years before any other appearance of the area 63 puzzle. This is dubious as Loyd did not claim this as his invention in other places where he was describing his accomplishments. In 1928, Sam Loyd Jr. [11] describes the area 63 version as something he had discovered, but makes no claims about the area 65 form, although he often claimed his father's inventions as his own. For example, on p. 5, he says "My 'Missing Chinaman Puzzle' " of 1896.

Let us look at the classic 8 × 8 to 5 × 13 of Figure 16.2. The history of this particular version is obscure. It is shown in Loyd's 1914 Cyclopedia [10] (p. 288 & 378). Loyd also gives the solution,

IX. Ein geometrisches Paradoxon. Um ad oculos zu demonstriren, dass das Schachbret nicht nur 64, sondern auch 65 Felder besitzt, schneide man dasselbe aus starkem Papier, zerlege es auf die in Fig. 1

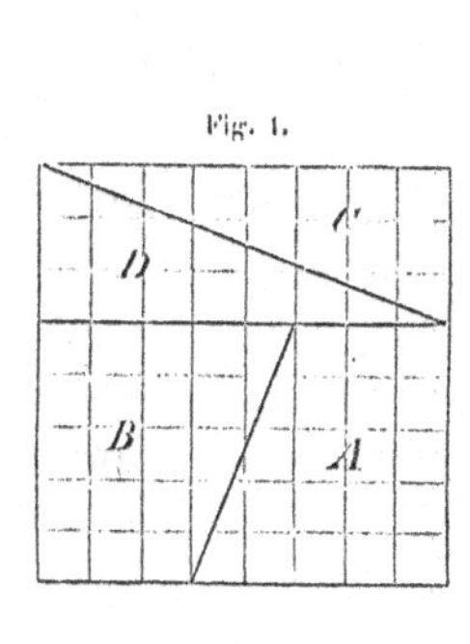

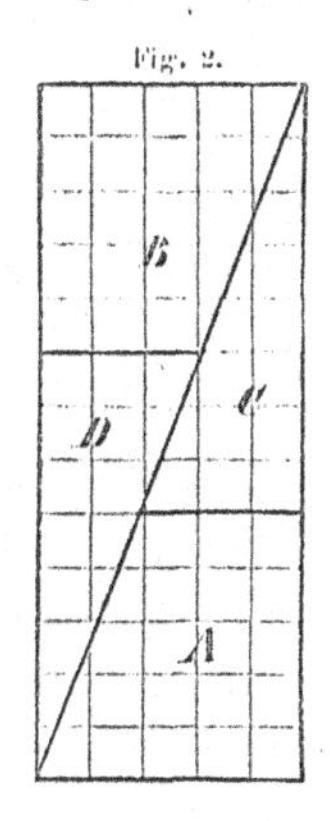

angegebene Weise in vier, zu je zweien congruente Stücke *A*, *B*, *C*, *D* und setze diese zu einem Rechtecke zusammen, welches, wie Fig. 2 zeigt, die Grundlinie 5 und die Höhe 13 besitzt also 65 Felder enthält. — Wir theilen diese kleine Neckerei mit, weil die Aufsuchung des begangenen Fehlers eine hübsche Schüleraufgabe bildet und weil sich an die Vermeidung des Fehlers die Lösung und Construction einer quadratischen Gleichung knüpfen lässt. Schl.

Figure 16.2. Schlömilch, 1868, [14].

IN PUZZLELAND

Tommy Riddles tells us that we need know nothing about checkers or chess to solve these puzzles. King Puzzlepate is trying to place the greatest number of men on a chess board without having three men in line in any possible direction. He has started by placing the first two men correctly; now it is up to you to assist him by adding as many men as possible without getting any three in line.

We are told that the first checkerboard ever constructed, which was made by a man by the name of Siesa, and is still preserved in the British Museum, is made of four pieces, as the one shown in the second puzzle. Now the four pieces of this board can be rearranged together so as to make three different puzzles: A square board of 64 squares, an oblong one of 65, or an odd-shaped one of but 63. It is said that Dr. Slasher won the championship by this marvelous coup of arranging the four pieces so as to reduce the board to 63 squares. See if you are able to do it. There has been so much discussion regarding this paradoxical problem that occasion is taken to say that Mr. Loyd presented it before the first American Chess Congress in 1858.

A Charade.

I am what I was, which is so much the worse,
I'm not what I was, but quite the reverse;
From morning till night I do nothing but fret,
And sigh to be what I never was yet.

A Charade.

My first, a substance hard and bright,
Is useful, morning, noon, and night;
My second, find it where you will,
Is of the same dimension still:
And by my whole, I often try,
Butchers' and grocers' honesty.

SAM LOYD'S STAGE PUZZLE

An English tourist in the wild and woolly West was informed that if he wished to walk on to Piketown the stage would only get there one mile ahead of him for although it would get to a certain wayhouse while you were walking four miles, it waits there 30 minutes, so you would catch up in time to ride on to Piketown if you wished. "But," as the host of the Pilgrim's hotel remarked, "from these facts there is a clever way of figuring out how to beat the stage by 15 minutes!" Can you tell how far it was from the hotel to Piketown?

288

Figure 16.3. Loyd, 1914, [10], pp. 288 & 378.

Figure 16.3. Unfortunately this seemed smudged. He shows that both rectangles have chessboard coloring, and he is the first to indicate this. However, the smudge is deliberate to conceal the fact that the coloring does NOT match up! Indeed, the corners of a 5×13 board should all be the same color, but two of them in the solution arise from adjacent corners of the 8×8 chessboard and have opposite colors! Loyd also poses the related problem of arranging the four pieces to make a figure of area 63, as shown in Figure 16.4.

The first known publication of the 8×8 to 5×13, Figure 16.2, is in 1868, in a German mathematical periodical, signed Schl. [14]. It has been suggested the author was Otto Schlömilch, and this seems right as he was a co-editor of the journal at the time. In 1953, Coxeter [3] said it was V. Schlegel, but he apparently confused this with another article on the problem by Schlegel. Schlömilch does not give any explanation for this "teaser", leaving it as a student exercise!

By 1886, a writer [13] says: "We suppose all the readers . . . know this old puzzle." In 1877 [4], it was recognized that the paradox is

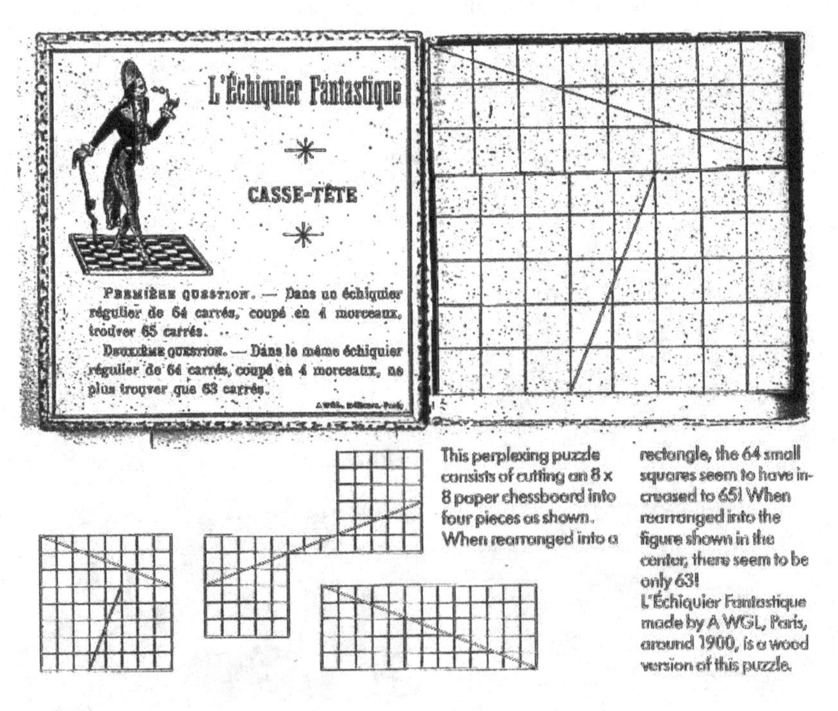

Figure 16.4. AWGL, c. 1900 [1 & 18].

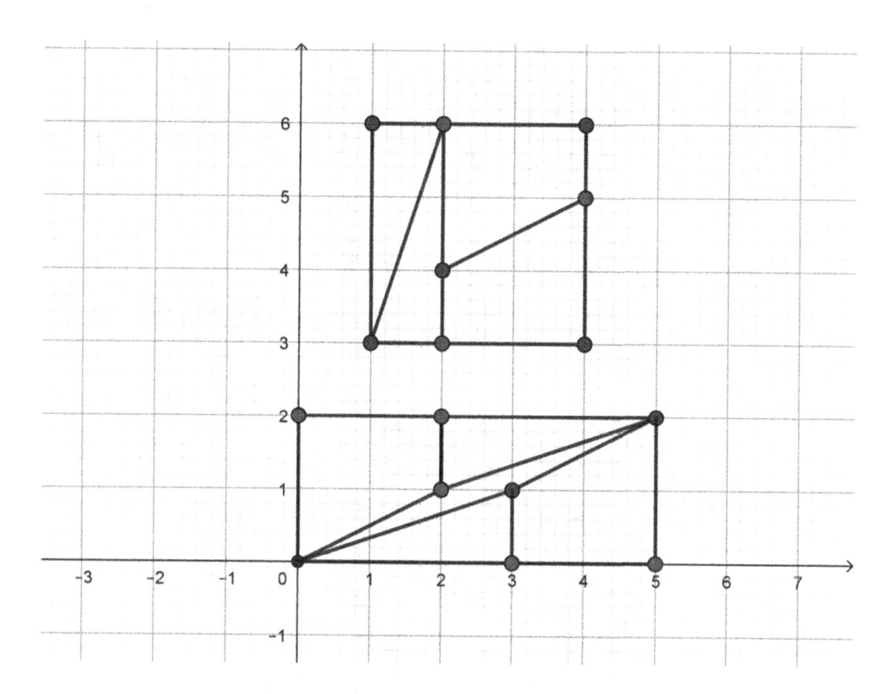

Figure 16.5. From 9 squares to 10 squares.

related to the fact that $5 \times 13 - 8 \times 8 = 1$ and that 5, 8, 13 are three consecutive Fibonacci numbers. Taking a smaller example based on the numbers 2, 3, 5 makes the trickery clear, see Figure 16.5. One can also make a 5×5 into a 3×8, but then there is a loss of area from the square form.

However, there are other versions of vanishing area or object puzzles. Since 1900 several dozen have been devised and there are examples where both some area and an object vanish!. See Figure 16.6.

There is a fairly common magician's trick, taking a 7×7 square array of £ signs and rearranging to get a 7×7 square array of £ signs and an extra unit square containing an extra *£* sign. One version, called "Credit Squeeze", using £, is attributed to Howard Gower, but Michael Tanoff kindly obtained for me an American version, using dollars, called "It's Magic DOLLAR DAZE", produced by Abbott's with no inventor named. Lennart Green uses a version of this in his magic shows, but he manages to reassemble it three times, getting an extra piece out each time! Needless to say, this involves further sleight of hand. A version of this is available on the Internet, under the name

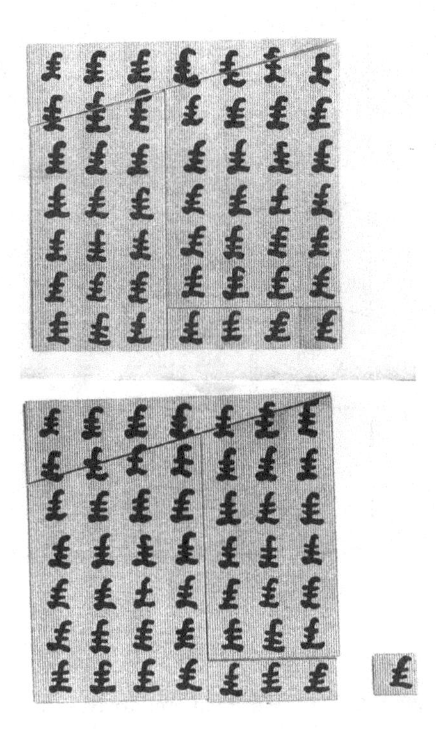

Figure 16.6. Credit Squeeze.

Missing Square Puzzle. (Note that this puzzle also incorporates the vanishing object aspect.)

16.1 Early Examples

Gardner and others tracked the idea back to Hooper [9] in 1783, as seen in Figure 16.7. Here we have a 3×10 cut into four pieces which make a 2×6 and a 4×5. However, Hooper's first edition of 1774 erroneously has a 3×6 instead of a 2×6 rectangle and notes there are now 38 units of area. This was corrected in the second edition of 1783 and this version occurs fairly regularly in the century following Hooper.

Gaspar Schott [15] contains a description of a 1537 version due to the 16th century architect Sebastiano Serlio [17]. The Serlio reference is in his famous treatise in five books on architecture, still in print in several languages.

286 RATIONAL RECREATIONS. 287

RECREATION CVI.

The geometric money.

DRAW on pafteboard the following rectangle ABCD, whofe fide AC is three inches, and AB ten inches.

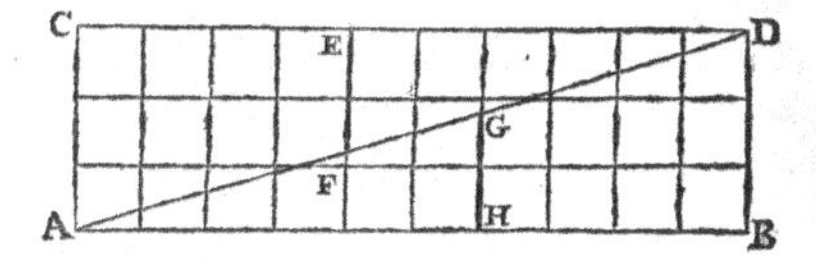

Divide the longeft fide into ten equal parts, and the fhorteft into three equal parts, and draw the perpendicular lines, as in the figure, which will divide it into thirty equal fquares.

From A to D draw the diagonal A D, and cut the figure, by that line, into two equal triangles, and cut thofe triangles into two parts, in the direction of the lines E F and G H. You will then have two triangles, and two four-fided irregular figures, which you are to place together, in the manner they ftood at firft, and in each fquare you are to draw the figure of a piece of money; obferving to make thofe in the fquares, through which the line AD paffes, fomething imperfect.

As the pieces ftand together in the foregoing figure, you will count thirty pieces of money only; but if the two triangles and the two irregular figures be joined together, as in the following figures, there will be thirty-two pieces.

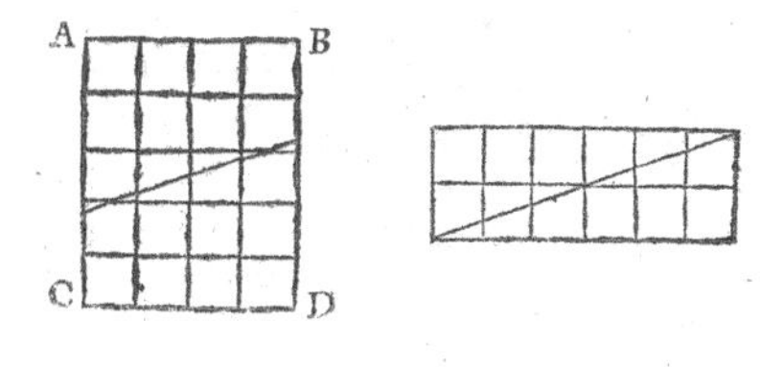

RECRE-

Figure 16.7. Hooper, 1774, [9].

Figure 16.8 [17], f. 12v. He is taking a 3×10 board, ABDC, and cuts it diagonally, then slides one piece by 3 to form an area 4×7 with two bits sticking out, which he then trims away. He doesn't notice that this implies that the two extra bits form a 1×3 rectangle and hence does not realize the change in area implied.

Already in 1567, Pietro Cataneo [2] pointed out the mistake and what the correct process would be. A later discussion of this in Schwenter [16], citing another architect, was well known in the 16th and 17th centuries, but the knowledge disappeared despite the fact that Serlio's book has been in print in Italian, Dutch, French and English since that time and Schwenter was fairly well-known.

There are two other late 18th century examples, possibly predating Hooper. Charles Vyse's *The Tutor's Guide, ...* [19] was a

Of Geometrie

It may also fall out, that a man should finde a Table of ten foote long, and three foote broade: with this Table a man would make a doore of seuen foote high, and foure foote wyde. Now to doe it, a man would saw the Table long wise in two parts, and setting them one vnder another, and so they would be but fiue foote high, and it should bee seuen: and againe, if they would cut it three foote shorter, and so make it foure foote broade, then the one side shall be too much peeced. Therefore he must doe it in this sort: Take the Table of ten foote long, and three foot broad, & marke it with A. B. C. D. then sawe it Diagonall wise, that is, from the corner C. to B. with two equall parts, then drawe the one peece thereof three foote backwards towards the corner B. then the line A. F. shall be foure foote broad, and so shall the line E.D. also hold foure foote broad: by this meanes you shall haue your doore A. E. F. D. seuen foote long, and foure foote broade, and you shall yet haue the three cornerd pieces marked E. B. G. and C. F. and C. left for some other vse.

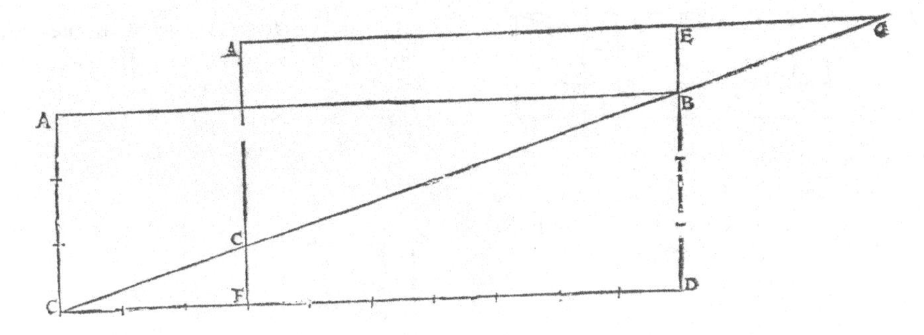

Figure 16.8. Serlio, later 1611 edition.

popular work, going through at least 16 editions, during 1770–1821. The problem is: "A Lady has a Dressing Table, each side of which is 27 Inches, but she is desirous to know how each Side of the same may = 36 Inches, by having 4 foot of Plank, superficial Measure, joined to the same. The Plan in what Manner the Plank must be cut and applied to the Table is required?" [The plank is one foot wide.] The solution is in *The Key to the Tutor's Guide; ...* (p. 358).

She cuts the board into two $12'' \times 24''$ rectangles and cuts each rectangle along a diagonal. By placing the diagonals of these pieces on the sides of her table, she makes a table $36''$ square. But the hypotenuses of these triangles are $12\sqrt{5} = 26.83\ldots''$ See Figure 16.9.

Note that $27^2 + 12 \cdot 24 = 1305$ while $36^2 = 1296$. Vyse is clearly unaware that area has been lost. By dividing all lengths by 3, one gets a version where one unit of area is lost. Note that 4, 8, 9 is almost a Pythagorean triple. It is likely to occur in the first edition of 1770 and hence predate Hooper. Like Serlio, the author is unaware that some area has vanished!

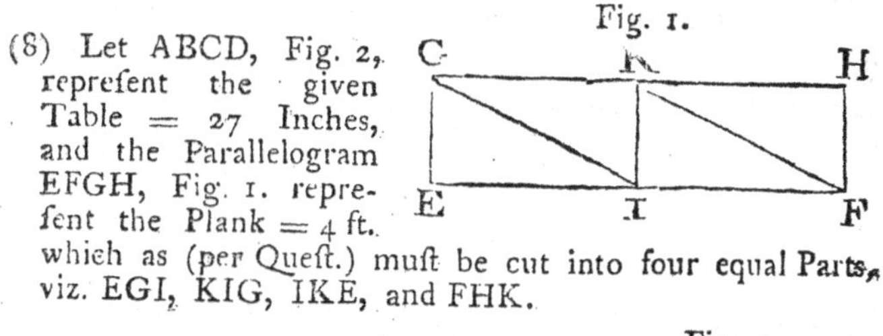

(8) Let ABCD, Fig. 2, represent the given Table = 27 Inches, and the Parallelogram EFGH, Fig. 1. represent the Plank = 4 ft. which as (per Quest.) must be cut into four equal Parts, viz. EGI, KIG, IKE, and FHK.

Then by properly applying the four equal Parts in the Parallelogram EFGH, to the given Table ABCD, then will EGI be =to NBD, KIG=LAB, IKF =MCA, and HFK will be= to ODC, which makes LMNO, a complete Square, each Side thereof=36 Inches, which was required.

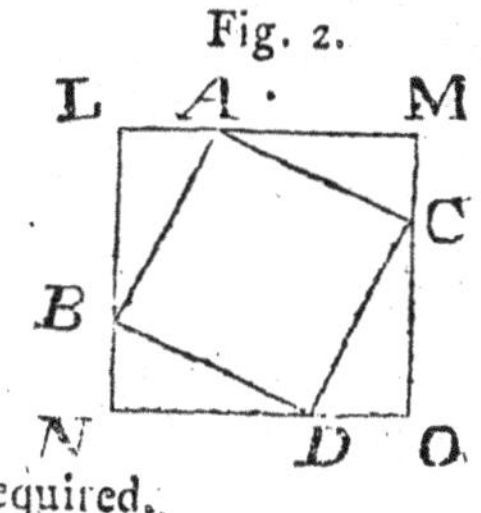

Figure 16.9. Vyse Solution, c. 1771.

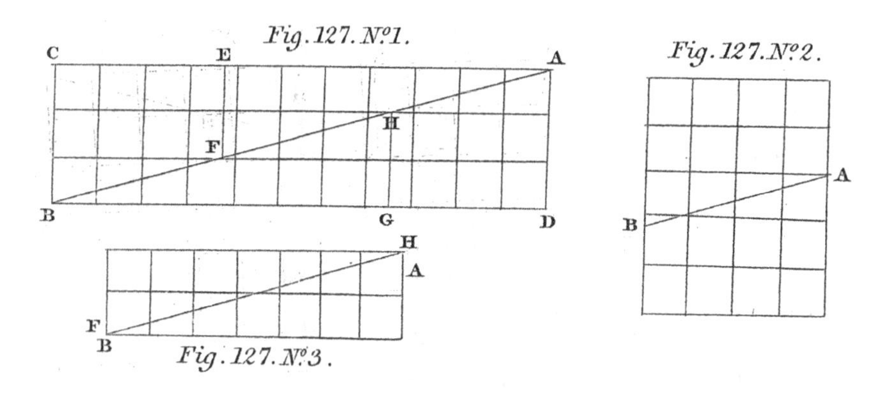

Figure 16.10. Ozanam, 1778, 1790 reissue.

The 1778 edition of Ozanam by Montucla [12] has an improvement on Hooper. Figure 16.10 is figure 127, plate 16, p. 363 for prob. 21, pp. 302–303 in the 1790 reissue. This has 3×11 to 2×7 and 4×5. Here just one unit of area is gained, instead of two units as in Hooper. He remarks that M. Ligier probably made some such mistake in showing $17^2 = 2 \cdot 12^2$ and this is discussed further on the later page.

In conclusion, we have found that vanishing area puzzles are at least two hundred years older than Martin Gardner had found. We have also found a number of new forms of the puzzle. Who knows what may turn up as we continue to examine old texts? Martin would have greatly enjoyed these results.

Bibliography

[1] AWGL [Alphonse Wogue and G. Levy]. *L'Echiquier Fantastique.* c. 1900. (Wooden puzzle of 8×8 to 5×13 and to area 63. See also Loyd's *Cyclopedia.* 144.)

[2] P. Cataneo. *L'Architettura di Pietro Cataneo Senese.* Aldus, 1567, 164–165.

[3] H. S. M. Coxeter. "The golden section, phyllotaxis, and Wythoff's game." *Scripta Mathematica,* 19 (1953) 135–143.

[4] G. H. Darwin. "A geometrical puzzle." *Messenger of Mathematics,* 6 (1877) 87.

[5] W. Dexter. "Some postcard puzzles." *Boy's Own Paper* (14 December 1901) 174–175.

[6] M. Gardner. "Geometrical Vanishes." in *Mathematics, Magic and Mystery.* Dover, 1956. Chaps. 7–8.

[7] M. Gardner. "The author pays his annual visit to Dr. Matrix, the numerologist." *Scientific American* (January 1963) 138–145. Reprinted in *The Numerology of Dr. Matrix,* Simon & Schuster, NY, 1967; in *The Incredible Dr. Matrix,* Simon & Schuster, NY, 1967; and in *The Magic Numbers of Dr. Matrix,* Prometheus, Buffalo, 1985; as Chap 3 in each of these.

[8] M. Gardner. Advertising premiums to beguile the mind: classics by Sam Loyd, master puzzle-poser. *Scientific American* (November 1971) 114–121. Reprinted in *Wheels, Life and Other Mathematical Amusements,* Freeman, 1983, Chap. 12, 124–133.

[9] HOOPER. 286–287: Recreation CVI—The geometric money. The image is taken from the 2nd ed.

[10] S. Loyd, ed. by Sam Loyd Jr. *Sam Loyd's Cyclopedia of 5,000 Puzzles, Tricks and Conundrums.* Lamb Publishing or Bigelow, NY, 1914 = Pinnacle or Corwin, NY, 1976, 323 (Get off the Earth); 288 & 378 (area 64 to areas 65 and 63 versions). (This is a reprint of Loyd's *Our Puzzle Magazine,* a quarterly which started in June 1907 and ran till 1908. From known issues, it appears that these problems would have appeared in October 1908 and June 1908.)

[11] S. Loyd Jr. " A paradoxical puzzle." *Sam Loyd and His Puzzles.* Barse & Co., 1928, 19–20 & 90.

[12] OZANAM.

[13] R. A. Proctor. "Some puzzles." *Knowledge,* 9 (August 1886) 305–306.

[14] Schl.[b] "Ein geometrisches Paradoxon." *Z. Math. Phys.* 13 (1868) 162.

[15] G. Schott. *Magia Universalis.* Joh. Martin Schönwetter, Bamberg, Vol. 3, 1677, 704–708.

[16] D. Schwenter. *Delici Physico-Mathematicae. Oder Mathemat-und Philosophische Erquickstunden,* .... Jeremi Dümlers, Nuremberg, 1636. Probably edited for the press by Georg Philip Harsdörffer. [Extended to three volumes by Harsdörffer in 1651 & 1653, with vol. 1 being a reprint of the 1636 vol.] 451.

[17] S. Serlio. *Libro Primo d'Architettura.* 1545. This is the first part of his Architettura series of 5 books, 1537–1547, first published together in 1584. (also Dover, 1982, reprint of the 1611 English edition.)

[18] J. Slocum and J. Botermans. *Puzzles Old & New—How to Make and Solve Them.* Univ. of Washington Press, 1986, 144.

[19] C. Vyse. "Problem 8." in *The Tutor's Guide,* .... circa 1771. (Only seen in 10th edition, by J. Warburton. S. Hamilton eds, G. G. and J. Robinson, 1799, 317.) (The solution is Figure 10, in *The Key to the Tutor's Guide,* .... 8th ed., 1802, 358.)

[b] Probably Otto Schlömilch.

Appendix A

Ancient and Important Sources

This appendix has three sections. The first gives the bibliographic entries that were previously merely abbreviated. The second describes the larger Sources project. And the third describes some problems that seem to need more research.

A.1 Bibliography of Early Work

The reason these citations were postponed until here is simply that the entries are not straightforward. When a citation is to an old manuscript or incunabula or material typically found in a rare book room things become tricky. In such cases there is rarely a canonical version of the text. In fact the history of the text can become intertwined with the material in the text. Multiple version are common. Different translators and commentators may disagree.

Abbot Albert Abbot Albert von Stade. *Annales Stadenses.* c. 1240. Ed. by J. M. Lappenberg. in *Monumenta Germaniae Historica*, Scriptorum t. XVI, Imp. Bibliopolii Aulici Hahniani, Hannover, 1859, 271–359, particularly pp. 332–335.

Vogel dates this as 1179, but Tropfke gives 1240, which is more in line with Lappenberg's notes on variants of the text. The material of mathematical interest (on 332–335), and several other miscellaneous sections, are inserted at the year 1152 of the *Annales*, so perhaps Vogel was misled. A translation of the problems is in Chapter 5.

Ahrens Wilhelm Ahrens. *Mathematische Unterhaltungen und Spiele.* 2 volumes, 2nd ed., 1910/1918, Teubner. Chapter 4 (The first edition was 1901 and the third was 1921.)

Alcuin Alcuin (of York). *Propositiones ad acuendos juvenes* [Problems to sharpen the young], c. 800. The definitive version is ALCUIN-FOLKERTS An English translation by John Hadley and David Singmaster appears in this volume as Chapter 4.

It has been attributed to both Alcuin and Bede. The material can be found under both names in the *Patrologiae Latinae*, t. 101, col. 1143–1160 and t. 90, col. 675–672, respectively. See BEDE. Folkerts had found 12 manuscripts of the Propositiones, the earliest being from the late 9th century. This oldest MS is not complete and the next oldest is missing some pages, but both contain the extra problems 11a, 11b and 33a of the Bede text. The next oldest MS is from late 10th century and is essentially the Alcuin text. Folkerts feels that the MSS fall into two groups or editions, roughly corresponding to the Alcuin and Bede texts, with the Bede representing a poor later edition of the Alcuin. He produces a reasonably definitive edition of the original, but it does not vary greatly from the Alcuin and Bede texts mentioned above. He uses the same numbers for the problems as the Alcuin text and numbers the extra Bede problems as 11a, 11b, 33a.

Alcuin (c. 735–804). *Propositiones Alcuini doctoris Caroli Magni Imperatoris ad acuendos juvenes.* 9th century. in B. Flacci Albini seu Alcuini, "Abbatis et Caroli Magni Imperatoris Magistri." *Beati Flacci Albini seu Alcuini, Abbatis et Caroli Magni Imperatoris Magistri. Opera Omnia: Operum pars octava: Opera dubia.* Edited by D. Frobenius, *Tomus secundus, volumen secundum.* Ratisbon, 1777, pp. 440–448. Revised by J. P. Migne as: *Patrologiae Cursus Completus: Patrologiae Latinae, Tomus 101*, Migne, 1863, columns 1143–1160. We call this the Alcuin text.

Alcuin–Folkerts Menso Folkerts. "Die älteste mathematische Aufgabensammlung in lateinischer Spräche: Die Alkuin zugeschriebenen Propositiones ad Acuendos Iuvenes; Denkschriften der Österreichischen Akademie der Wissenschaften," *Mathematische naturwissenschaftliche Klasse* 116 (6) (1978) 13–80. (Also separately published by Springer, Vienna, 1978.) The critical part is somewhat revised as: "Die Alkuin zugeschriebenen 'Propositiones ad Acuendos Iuvenes,"' in BUTZER, pp. 273–281.

Alcuin–Folkerts–Gericke Menso Folkerts & Helmuth Gericke. "Die Alkuin zugeschriebenen Propositiones ad Acuendos Juvenes (Aufgaben zur Schärfung des Geistes der Jugend)." in BUTZER, pp. 283–362.

Al-Karagi Aboû Beqr Mohammed Ben Alhaçen Alkarkhî [= al Karagi]. Untitled MS called Alfakhrî. c. 1010. MS 952, Supp. Arabe de la Bibliothèque Impériale, Paris. Edited into French by Franz Woepcke as *Extrait du Fakhrî*, L'Imprimerie Impériale, Paris, 1853. (also Georg Olms Verlag, 1982.)

Āryabhaṭa Āryabhaṭa (I) (476–550). *Āryabhaṭīya*. 499.
Critically edited and translated into English by Kripa Shankar Shukla, with K. V. Sarma. Indian National Science Academy, 1976. (Volume 1 of a three volume series — the other two volumes are commentaries, of which Vol. 2 includes the commentary Āryabhaṭīya-Bhāṣya, in 629, see BHĀSKARA I . The other commentaries are later and of less interest.) (There is an older translation by Walter Eugene Clark: *The Āryabhavīya of Āryabhaṭa*. University of Chicago Press, 1930.)

Aslam Abū Kāmil Shujā ibn Aslam. *Kitāb al-ṭaraif fil ḥisāb* [Book of Rare Things in the Art of Calculation]. c. 900. Trans. by H. Suter, "Das Buch der Seltenheiten der Rechenkunst von Abū Kāmil el Miṣrī," *Bibl. Math.* (3) 11 (1910–1911) 100–120. (Part is in English in Ore, *Number Theory and Its History*, McGraw-Hill, NY, 1948, pp. 139–140.)

Bachet Claude-Gaspar Bachet. *Problèmes plaisans et délectables qui se font par les nombres.* 2nd ed., P. Rigaud, Lyon, 1624. (1st ed., 1612) The 5th ed., edited by A. Labosne in 1884 (Gauthier Villars) added material. (5th ed. reprinted by Blanchard, 1959.) Later editions use *plaisants.*

Ball W. W. Rouse Ball. *Mathematical Recreations and Essays.* Editions 1–11, Macmillan, London, 1892–1939; edition 12, Univ. of Toronto Press, Toronto, 1974.; edition 13 (with material by H. S. M. Coxeter), Dover, NY, 1987. The first three editions were entitled *Mathematical Recreations and Problems.*

Bakhshālī The Bakhshālī Manuscript, c. 7th century. This MS was found in May 1881 near the village of Bakhshālī, in the Yusufzāī district of the Peshawar division, then in India, but now in Pakistan. David Pingree says it is 10th century, but his student Hayashi opts for c. 7th century which seems pretty reasonable.

Bede Venerable Bede. Text published with and confused with Alcuin. "De Arithmeticis Propositiones." in *Opera Omnia: Pars Prima, Sectio II Dubia et Spuria* Tomus 1. Joannes Herwagen, 1563. Columns 135–146. Revised by J. P. Migne as: *Patrologiae Cursus Completus: Patrologiae Latinae, Tomus 90*, Migne, 1904, columns 665–676. *Incipiunt aliae propositiones ad acuendos juvenes* is col. 667–676. We call this the Bede text.

Many of the MSS have the *Aliae Propositiones ad Acuendos Juvenes* (Alcuin) following *De Arithmeticis Propositiones* which was usually attributed to Bede, so both were often attributed to Bede. However, in common with other works in the 1563 volume, the authorship of both works is generally considered spurious, so *De Arithmeticis Propositiones* is now ascribed to a Pseudo-Bede.

Bhāskara I Bhāskara. *Āryabhaṭīya-Bhāṣya.* 629. Critically edited by Kripa Shankar Shukla, Indian National Science Academy, New Delhi, 1976, including an English Appendix of the numerical examples used. (Volume 2 of a three volume series devoted to the Āryabhaṭīya (499) of Āryabhaṭa. Bhāskara I repeats and exposits Āryabhaṭa verse by verse, but Āryabhaṭa rarely gives numerical examples, so Bhāskara I provided them and these were later used by other Indian writers such as Chaturveda, 860.

Bhāskara II Bhāskara, *Bîjagaṇita,* 1150. in *Algebra, with Arithmetic and Mensuration from the Sanscrit of Brahmegupta and Bháscara,* translation by Henry Thomas Colebrooke, John Murray, 1817. (Also includes Bhāskara, *Lîlâvatî.* (Also Sändig, 1973.)

Brahmagupta Brahmagupta. *Bráhma-sphuṭa-siddhânta of Brahmagupta,* 628. in *Algebra, with Arithmetic and Mensuration from the Sanscrit of Brahmegupta and Bháscara,* translation by Henry Thomas Colebrooke, John Murray, 1817. (Also includes Bhāskara, *Lîlâvatî.*) (Also Sändig, 1973.) It appears from Colebrooke (p. v) and Datta (p. 10) that almost all the illustrative examples and all the solutions are due to Chaturveda Pṛthudakasvâmî in 860. Brahmagupta's rules are sometimes so general that one would not recognize their relevance.

Butzer P. L. Butzer, H. Th. Jongen and W. Oberschelp (eds). *Charlemagne and his Heritage 1200 Years of Civilization and Science in Europe: Vol. 2 Mathematical Arts*, Brepols, 1998.

Calandri Filippo Calandri. *Aritmetica.* c. 1500. Italian MS in Codex 2669, Biblioteca Riccardiana di Firenze. Edited by Gino Arrighi, Edizioni Cassa di Risparmio di Firenze, Florence, 1969. Volume 1 is a facsimile of the MS. Volume 2 is a transcription of the text — copies not in a set have 8 color plates inserted. F. 98v of the original is reproduced in color opp. p. 144 in independent copies of vol. 2 and its text is given on p. 196.

Cardan I Jerome Cardan, [= Girolamo Cardano = Hieronymus Cardanus] (1501–1576). *Practica Arithmetice, & Mensurandi Singularis.* Bernardini Calusci, 1539. Included in *Opera Omnia*, Joannis Antonius Huguetan & Marcus Antoniius Ravaud, Lyon, 1663 (often reprinted, e.g. in 1967) vol. IV. I will give the folios from the 1539 edition followed by the pages of the 1663 edition, e.g. ff. T.iiii.v–T.v.r (p. 113).

Cardan II Jerome Cardan. *Practica Arithmeticae Generalis.* Bernardini Calusci, 1539. Included in *Opera Omnia*, Joannis Antonius Huguetan & Marcus Antoniius Ravaud, Lyon, 1663 (often reprinted, e.g. in 1967), vol. IV.

Cardan III Jerome Cardan. *De Rerum Varietate.* Henricus Petrus, Basel, 1557; 2nd ed., 1557; 5th ed., 1581. Included in *Opera Omnia*, Joannis Antonius Huguetan & Marcus Antoniius Ravaud, Lyon, 1663 (often reprinted, e.g. in 1967), vol. III.

Cardan IV Jerome Cardan. *De Subtilitate Libri XXI*; J. Petreium, Nuremberg, 1550, and many later printings and editions; Liber XV; *Instrumentum ludicrum*, pp. 294–295. This is a very cryptic description.

Chiu Anonymous. *Chiu Chang Suan Ching = Jiu Zhang Suan Shu* [Nine Chapters on the Mathematical Art]. Variously dated from 150

BCE to 100 CE. German translation by K. Vogel, *Neun Bücher arithmetischer Technik*, Vieweg, 1968. My citations are to the pages in Vogel.

Chuquet Nicolas Chuquet. *Problèmes numériques faisant suite et servant d'application au Triparty en la science des nombres de Nicolas Chuquet Parisien.* MS no. 1346 du Fonds Français de la Bibliothèque Nationale, 1484, ff. 148r–210r. Abridged: Aristide Marre. "Appendice au Triparty en la science des nombres de Nicolas Chuquet Parisien." *Bulletino di bibliografia e di storia delle scienze matematiche e fisiche*, 14 (1881) 413–460.

Columbia Algorism Anonymous. *The Columbia Algorism.* Italian MS of c. 1370 at Columbia University as X511 A1 3. Transcribed and edited by K. Vogel: *Ein italienisches Rechenbuch aus dem 14. Jahrhundert.* (Veröffentlichungen des Forschungsinstituts des Deutschen Museums für die Gesch. der Naturw. und der Tech., Reihe C: Quellentexte und Übersetzungen, Nr. 33), 1977. My page references will be to this edition. (See also: Elizabeth B. Cowley. "An Italian mathematical manuscript." In *Vassar Medieval Studies*, edited by Christabel Forsyth Fiske, Yale University Press, 1923, 379–405 — especially 393, 402–403 and the plates between 402 and 403.)

dell'Abbaco-Aritmetica Paolo dell'Abbaco (sometimes called Dagomari) (c. 1281–1370). *Trattato d'Aritmetica.* c. 1370. See *Codex Magliabechiano XI, 86 at Biblioteca Nazionale di Firenze.* Edited by Gino Arrighi, Domus Galilaeana, Pisa, 1964.

Warren Van Egmond, "New light on Paolo dell'Abbaco," *Annali dell'Istituto e Museo di Storia della Scienza di Firenze* 2 (2) (1977) 3–21, asserts this is a later compilation and doubts that it is due to dell'Abbaco, preferring "pseudo-dell'Abbaco, 1440."

dell'Abbaco-Tutta Paolo Dagomari (dell'Abbaco). *Trattato di tutta l'arte dell'abacho.* c. 1339. Codex 167 in Plimpton collection, Columbia Univ. D. E. Smith; *Rara Arithmetica*, 4th ed., with De Morgan's *Arithmetical Books*; Chelsea, 1970. This is quite a different book than *Trattato d'Aritmetica.* It includes the *Regoluzze* which is sometimes cited separately.

Diophantos Diophantus. *Arithmetica.* c. 250. in T. L. Heath. *Diophantus of Alexandria* 2nd ed. OUP, 1910. (also Dover, 1964.)

Euler Leonard Euler. *Vollständig Anleitung zur Algebra.* Royal Academy of Sciences at Petersburg, 1770. [A Russian translation had appeared in 1768.] Translated into French by John III Bernoulli in 1774. Translated from French into English, with further notes, by Rev. John Hewlett, with a "Memoir of Euler" by Francis Horner [Horner actually did the translation; Hewlett edited it]: *Elements of Algebra*, 1797. (also Springer, 1984.)

Fibonacci Leonardo Pisano, called Fibonacci. (c. 1170–1240) *Liber Abbaci.* 1202 (2nd edition, 1228). [The dates of 1202 and 1228 are based on the Pisan calendar.] in *Scritti di Leonardo Pisano*; vol. I. Edited by B. Boncompagni, Tipografia delle Scienze Matematiche e Fisiche, Rome, 1857. Translated by Laurence E. Sigler as *Fibonacci's Liber Abaci: A Translation into Modern English of Leonardo Pisano's Book of Calculation.* Springer, New York, 2002. I have added page references to this, denoted S, after the Boncompagni pages, e.g. pp. 397–398. (S: 543–544) [The title pages give "abbaci," but the text begins "Incipit liber Abaci ... Anno MCCII." while a c. 1275 MS starts "Incipit abbacus." Both forms are used, sometimes even in the same article.] [In September 1994 and March 1998, I examined Siena L.IV.20 and 21 and Conv. Soppr. C.1.2616.]

Gardner Martin Gardner. *Mathematical Games: The Entire Collection of His Scientific American Columns.* CD-ROM. Mathematical Association of America. A revised collection from decades of columns and 15 book anthologies.

Gerhardt Frater Friedrich Gerhardt (attributed). *Algorismus Ratisbonensis.* Latin & German MSS, c. 1450. Transcribed and edited from 6 MSS: Kurt Vogel. *Die Practica des Algorismus Ratisbonensis.* C. H. Beck'sche Verlagsbuchhandlung, 1954. (Vogel says (on p. 206) that almost all of Munich 14684 is included in this.)

Gori Dionigi Gori. *Libro di arimetricha.* 1571. Italian MS in Biblioteca Comunale di Siena, L. IV. 23. Extensively quoted and discussed in: R. Franci & L. Toti Rigatelli. *Introduzione all'Aritmetica Mercantile del Medioevo e del Rinascimento.* Istituto di Matematica dell'Università di Siena, [1980?]. (The "Introduzione" was later published by Quattroventi, Urbino, 1981.)

Hooper William Hooper. *Rational Recreations, In which the Principles of Numbers and Natural Philosophy Are clearly and copiously* *elucidated, by a series of Easy, Entertaining, Interesting Experiments. Among which are All those commonly performed with the cards,* 4 vols. L. Davis *et al.*, 1774. (2nd ed., corrected, 1783–1782; 3rd ed., corrected, 1787; 4th ed., corrected, B. Law *et al.*, London, 1794.)

Jackson John Jackson. *Rational Amusement for Winter Evenings; or, A Collection of above 200 Curious and Interesting Puzzles and Paradoxes relating to Arithmetic, Geometry, Geography, & c. With Their Solutions, and Four Plates. Designed Chiefly for Young Persons*; London: Sold by J. and A. Arch, Cornhill; and by Barry & Son, High-Street; and P. Rose; Bristol; 1821.

Loyd Sam Loyd. *Sam Loyd's Cyclopedia of 5,000 Puzzles, Tricks and Conundrums.* Lamb Publishing, 1914. (also Pinnacle or Corwin, 1976.) Actually edited by Sam Loyd II. [This book is essentially a reprint of Loyd's *Our Puzzle Magazine*, a quarterly which started in June 1907.]

Lucas Édouard Lucas. *L'Arithmétique Amusante.* Edited by H. Delannoy, C.-A. Laisant and E. Lemoine, Gauthier-Villars, 1895. (also Blanchard,1974.)

Magician's Own Book Anonymous. *The Magician's Own Book, or The Whole Art of Conjuring. Being a Complete Hand-Book of Parlor Magic, and Containing over One Thousand Optical, Chemical, Mechanical, Magnetical, and Magical Experiments, Amusing Transmutations, Astonishing Sleights and Subtleties, Celebrated Card Deceptions, Ingenious Tricks with Numbers, Curious and Entertaining Puzzles, Together with All the Most Noted Tricks of Modern Performers. The Whole Illustrated with over 500 Wood Cuts, and Intended as a Source of Amusement for One Thousand and One Evenings.* Dick and Fitzgerald, 1857. [The authorship is a matter of debate. Henry Llewellyn Williams Jr. is generally credited with it, probably assisted by Mr. Dick and John Wyman. There was a UK book of the same title but quite different content and there were several books which used large amounts of this book.]

Mahāvīrā(chārya) Mahāvīrā(chārya). *Gaṇita sāra sangraha* (= Gaṇita sāra saṁgraha = Ganitasar Samgrha). 850. Translated by M. Rangāchārya. Government Press, Madras, 1912. The sections in this are verses. We will refer to the integral part of the first verse of the problem, so where he uses 121.5–123, we will use v. 121. This work is described by David Eugene Smith; "The Ganita-Sara-Sangraha of Mahāvīrāchārya"; *Bibliotheca Mathematica*, 3 (1908/09) 106–111.

Sanskrit text with English transliteration, Kannada translation and notes by Dr. (Mrs.) Padmavathamma; English translation and notes by Rao Bahadur M. Rangāchārya (from the 1912 version). Sri Siddhāntakīrthi Granthamāla, Sri Hombuja Jain Math, Hombuja, 577 436, Karnataka, 2000.

Metrodorus Metrodorus. *The Greek Anthology.* c. 510. Translated by W. R. Paton as part of the Loeb Classical Library, Harvard University Press (also Heinemann), 1916–1918. Vol. 5, Book 14. [This contains 44 mathematical problems (Art. 145 & 146, pp. 104–105) most of which are attributed to Metrodorus, though he is clearly simply a compiler.]

Minguét Pablo Minguét è (or é) Yról (or Irol). *Engaños à Ojos Vistas, y Diversion de Trabajos Mundanos, Fundada en Licitos Juegos de Manos, que contiene todas las diferencias de los Cubiletes, y otras habilidades muy curiosas, demostradas con diferentes Láminas, para que los pueda hacer facilmente qualquier entretenido.* Pedro Joseph Alonso y Padilla, 1733. [This had a number of editions and printings: 1755, c. 1760, 1766, 1820, 1822, 1847, 1888, 1888 and there was a 1981 facsimile of an 1864 ed. The early history of this book is confused.] [The initial 25 pages is a fairly direct translation of the 1725 Ozanam, and a number of other pictures and texts also are taken from Ozanam.]

Munich Anonymous. *Munich Codex Lat. 14684.* 14th century. Ff. 30–33. Published as: M. Curtze. "Mathematisch-historische Miscellan: 6 — Arithmetische Scherzaufgaben aus dem 14 Jahrhundert." *Bibliotheca Math.* 2 (9) (1895) 77–88. [This has 34 problems and Curtze gives brief notes in German. Curtze says these problems also appear in *Codex Amplonianus* Qu. 345, ff. 16–16′, c. 1325. Later Curtze claimed this Codex is 13th century.] This comes from the same monastery (St. Emmeran) as Gerhardt, see GERHARDT.

Ozanam [The bibliography of this book is a little complicated. There are 19 (or 20) French and 10 English editions, from 1694 to 1854, as well as 15 related versions; see the next entries. The book first appeared in 1694, but the material of interest to us first appears in the new edition of 1723–1725 which extended the work from two volumes was printed with varying dates. Also the "privilege date" which predates publication has caused errors that are often repeated].

Jacques Ozanam. *Recreations Mathematiques et Physiques, qui contiennent Plusieurs Problémes* [*sic*] *utiles & agreables, d'Arithmetique, de Geometrie, d'Optique, de Gnomonique, de Cosmographie, de Mecanique, de Pyrotechnie, & de Physique. Avec un Traité nouveau des Horloges Elementaires.* 2 vols., Claude Jombert, Paris, 1694. [The Traité des Horologes élémentaires, is a translation of Domenico Martinelli's *Horologi elementari.*] There were ten reprints, some as one volume.

Ozanam-Grandin *Recreations Mathematiques et Physiques, qui contiennent Plusieurs Problêmes d'Arithmétique, de Géométrie, de Musique, d'Optique, de Gnomonique, de Cosmographie, de Mécanique, de Pyrotechnie, & de Physique. Avec un Traité des Horloges Elementaires. Nouvelle edition, Revûë, corrigée & augmentée.* 4 volumes. Claude Jombert, Paris, 1725.[a] "The editor is said to be one Grandin." It was reprinted often until 1770.

Ozanam-Montucla *Récréations Mathématiques et Physiques, Qui contiennent les Problêmes & les Questions les plus remarquables, & les plus propres à piquer la curiosité, tant des Mathématiques que de la Physique; le tout traité d'une maniere à la portée des Lecteurs qui ont seulement quelques connoissances légeres de ces Sciences. Par feu M. Ozanam, de l'Académie royale des Sciences, & c. Nouvelle Edition, totalement refondue & considérablement augmentée par M. de C. G. F.* Claude Antoine Jombert, fils aîné, 1778, 4 vols. (also Firmin Didot, 1790.) Jean Étienne Montucla revised this, under the pseudonym M. de C. G. F. (M. de Chanla, géométre forézien). This is a considerable reworking of the earlier versions. In particular, the interesting material on conjuring and mechanical puzzles in volume IV has been omitted.

[a]However, vol. 4 has a different title page. *Récréations Mathématiques et Physiques, ou l'on traite Des Phosphores Naturels & Artificiels, & des Lampes Perpetuelles. Dissertation Physique & Chimique. Avec l'Explication des Tours de Gibeciere, de Gobelets, & autres récréatifs & divertissans.*

Ozanam-Hutton *Recreations in Mathematics and Natural Philosophy: ...first composed by M. Ozanam, of the Royal Academy of Sciences, & c. Lately recomposed, and greatly enlarged, in a new Edition, by the celebrated M. Montucla. And now translated into English, and improved with many Additions and Observations, by Charles Hutton.* 4 volumes, T. Davison for G. Kearsley, 1803 (also Hurst, Rees, Orme and Brown, 1814.)

Ozanam-Riddle *Recreations in mathematics and natural philosophy: translated from Montucla's edition of Ozanam, by Charles Hutton, LL.D. F.R.S. & c. A new and revised edition, with numerous additions, and illustrated with upwards of four hundred woodcuts.* Thomas Tegg, 1840. Revised by Edward Riddle.

Pacioli-Summa Luca Pacioli (or Paciuolo). *Sūma de Arithmetica Geometria Proportioni & Proportionalita.* Paganino de Paganinis, Venice, 1494. A facsimile printed by Istituto Poligrafico e Zecca dello Stato, for Fondazione Piero della Francesca, Comune di Sansepolcro, 1994, with descriptive booklet edited by Enrico Giusti.

Pacioli-Divina Luca Pacioli (or Paciuolo). *De Divina Proportione* MS beginning: *Tavola dela presente opera e utilissimo compendio detto dela divina proportione dele mathematici discipline e lecto.* Three copies of this MS were made. One is in the Civic Library of Geneva, one is the Biblioteca Ambrosiana in Milan and the third is lost. Modern facsimiles exist. Contains illustrations of six Archimedean Polyhedra, and the first Stella Octangula.

Pacioli-DVQ Luca Pacioli (or Paciuolo). *De Viribus Quantitatis.* c. 1500. Italian MS in Codex 250, Biblioteca Universitaria di Bologna. A transcription by Maria Garlaschi Peirani, with assistance by Augusto Marinoni, has been published : Ente Raccolta Vinciana, Milano, 1997. Part 1: "Delle forze numerali cioe di Arithmetica" is extensively described in: A. Agostini Il. " 'De viribus quantitatis' di Luca Pacioli." *Periodico di Matematiche* 4 (4) (1924) 165–192. I will cite problems by number.

1509 Pacioli & da Vinci: [*De*] *Divina proportione Opera a tutti glingegni perspicaci e curiosi necessaria Ove ciascun studioso di Philosophia: Prospectiva Pictura Sculptura: Architectura: Musica: e altre Mathematice: suavissima: sottile: e admirabile doctrina consequira: e delectarassi: cōvarie questione de secretissima scientia.* Ill.

by Leonardo da Vinci. Including Piero della Francesca's *Libellus* and other extra material. Paganino de Paganini, Brescia, 1509. Modern facsimiles exist.

The printed version was assembled from three codices dating from 1497–1498 and contains the 1498 MS with several additional items. Pacioli & da Vinci give six Archimedean solids. They assert that the rhombi-cuboctahedron arises by truncating a cuboctahedron, but this is not exactly correct.

Pacioli wrote much of *De Viribus Quantitatis* in Milan. The printed version of the book included a version of PIERO and the handsome and often reproduced geometric designs for letters of the alphabet. Part of the printed version is *Libellus in tres partiales tractatus divisus quae corpori regularium e depēdentiū actine perscrutatiōis* ..., which is an Italian translation (probably by Luca Pacioli) of PIERO. Some architectural material and the handsome and often reproduced geometric designs for letters of the alphabet are also appended.

Davis says the drawings were made from models prepared by Da Vinci, but Pacioli made or had made at least three sets of 60 models.

Pearson A. Cyril Pearson. *The Twentieth Century Standard Puzzle Book.* Routledge, [1907]. Three parts in one volume, separately paginated.

Piero Piero della Francesca. *Libellus de Quinque Corporibus Regularibus.* c. 1487.

Piero would have written this in Italian and it is believed to have been translated into Latin by Matteo da Borgo, who improved the style. First post-classical discussion of the Archimedean polyhedra, but it was not published until an Italian translation (probably by Pacioli) was printed in Pacioli & da Vinci, qv, in 1509, as: *Libellus in tres partiales tractatus divisus quae corpori regularium e depēdentiū actine perscrutatiōis* ..., ff. 1–27. A Latin version was discovered by J. Dennistoun, c. 1850, and rediscovered by Max Jordan, 1880, in the Urbino manuscripts in the Vatican — MS Vat.Urb. lat. 632.

Davis identifies 139 problems in this, of which 85 (= 61%) are taken from the Trattato. The Latin text differs a bit from the Italian. Piero describes a sphere divided into 6 zones and 12 sectors. He gives truncated tetrahedron, truncated cube, truncated octahedron, truncated dodecahedron, truncated icosahedron — see below for the cubo-octahedron — and there is an excellent picture of the truncated tetrahedron on f. 22v of the printed version. The Latin MS gives different diagrams than in the 1509 printed version, including clear pictures of the truncated icosahedron and the truncated dodecahedron. J. V. Field showed that the counting is confused by the presence of the cubo-octahedron in the Trattato but not in the Libellus. So della Francesca rediscovered six Archimedean polyhedra, but only five appear in the *Libellus.* (See: Margaret Daly Davis. *Piero della Francesca's Mathematical Treatises The "Trattato d'abaco" and "Libellus de quinque corporibus regularibus."* Longo Editore, Ravenna, 1977.)

Prévost. J. Prévost. *La Première Partie des Subtiles et Plaisantes Inventions, Contenant Plusieurs Jeux de Récréation.* Antoine Bastide, 1584. Translated by Sharon King as: *Clever and Pleasant Inventions: Part One Containing Numerous Games of Recreations and Feats of Agility, by Which One May Discover the Trickery of Jugglers and Charlatans.* Hermetic Press, 1998. [No Part Two ever appeared.] This is apparently the first book primarily devoted to conjuring. A facsimile circulated in 1987.

The Rhind Papyrus. c 1650. Arnold Buffum Chace, ed. (1927–1929). Revised, National Council of Teachers of Mathematics, 1978. pp. 36–38. Also in: Gay Robins and Charles Shute. *The Rhind Mathematical Papyrus*; British Museum Publications, 1990.

Schwenter Daniel Schwenter (1585–1636). *Deliciae Physico Mathematicae. Oder Mathemat und Philosophische Erquickstunden, Darinnen Sechshundert Drey und Sechsig, Schöne, Liebliche und Annehmliche Kunststücklein, Auffgaben und Fragen, auf; der Rechenkunst, Landtmessen, Perspctiv, Naturkündigung und andern Wissenschafften genomën, begriffen seindt, Wiesolche uf der andern seiten dieses blats ordentlich nacheinander verzeichnet worden: Allen Kunstliebenden zu Ehren, Nutz, Ergössung des Gemüths und sonderbahren Wolgefallen am tag gegeben Durch M. Danielem Schwenterum.* Jeremiæ Dümlers, 1636. Extended to three volumes by Georg Philip Harsdörffer in 1651 and 1653, with the first being a reprint of the 1636 volume. There is a modern facsimile of the latter version edited by Jörg Jochen Berns, Keip Verlag, 1991.

Singmaster David Singmaster. "The history of some of Alcuin's *Propositiones*." in BUTZER, pp. 11–29. [Presented at: Colloquium Carolus Magnus — 1200 Jahre Wissenschaft in Zentral-Europa, Aachen, March 1995.]

Ṭabarī Moḥammed ibn Ayyūb Ṭabarī. *Miftāh al-mu'āmālāt.* c. 1075. Ed. by Moḥammed Amin Riyāḥi, Teheran, 1970.[b]

Tartaglia Nicolo Tartaglia.[c] (*La Prima Parte del*) *General Trattato di Numeri et Misure. Curtio Troiano.* Venice, 1556. In 1578, Guillaume Gosselin produced an annotated translation of parts 1 & 2 into French: *L'Arithmetique de Nicolas Tartaglia.* Also *Tutte l'Opere d'Arithmetica,* All'Insegna del Leone, 1592-1593, 2 vol.

van Etten Henrik van Etten. *Recreation Mathematique.* Hanzelet, 1624. Many later editions in several languages. The English edition, entitled *Mathematicall Recreations* ..., of 1633 and later, was published by William Leake with an appendix by William Oughtred.

Wecker John [Johann (or Hanss) Jacob] Wecker. *Eighteen Books of the Secrets of Art & Nature Being the Summe and Substance of Naturall Philosophy, Methodically Digested* .... (As: De Secretis Libri XVII; P. Perna, Basel, 1582 (?? not yet seen). Now much Augmented and Inlarged by Dr. R. Read. Simon Miller, London, 1660, 1661. (also Robert Stockwell, [c. 1988].)

Witgeest-I Simon Witgeest. *Het Nieuw Toneel der Konsten, Bestaande uyt Sesderley Stukken: het eerste, handelt van alderley aardige Speeltjes en Klugjes: het tweede, van de Verligt-konst in 't Verwen en Schilderen: het derde, van het Etzen en Plaat-shijden: het vierde, van de Glas-konst: het vijfde, heest eenige aardige remedien tegen alderley Ziekten: het sesde, is van de Vuur-werken. Uyt verscheyde Autheuren by een vergadert, door S. Witgeest, Middel-borger.* Jan ten Hoorn, 1679. Facsimile with epilogue by John Landwehr, A. W. Sijthoff's Uitgeversmaatschappij N. V., 1967. There were many editions, but Nanco Bordewijk has examined these and discovered that the 3rd ed. of 1686 was so extensively revised and extended as to constitute a new book, WITGEEST-II.

[b]Not read — described in Hermelink.
[c]Tartaglia's given name is now usually given as Niccol, but the book gives Nicolò.

Witgeest-II Simon Witgeest. *Het Natuurlyk Tover-Boek, Of 't Nieuw Speel-Toneel Der Konsten. Verhandelende over de agt hondert natuurlijke Tover-Konsten. so uyt de de Gogel-tas, als Kaartspelen, Mathematische Konsten, en meer andered diergelijke aerdigheden, die tot vermaek, en tijtkorting verstrecken. Mitsgaders een Tractaet van alderley Waterverwen, en verligteryen; Als oock Een verhandelinge van veelderley Blanketsels. Om verscheyde wel-ruykende Wateren, Poederen en Balsemen, als ook kostelijke beeydselen, om het Aensicht, Hals en Handen, wit en sagt te maecken, door Simon Witgeest.* Jan ten Hoorn, 1686. See WITGEEST-I. The new material is said to already be in the 2nd ed. of 1682. Translated into German as: *Naturliches Zauber-Buch oder neuer Spiel-Platz der Künste*; Hoffmanns sel Wittw. & Engelbert Streck, 1702, with many later editions, all apparently based on the 1682 edition. Landwehr has written a bibliographical article on this book.

Zeiller Zeiller, Martin. "Topographia Electorat, Brandenburgici et Ducatus Pomeraniae. uc. das ist Beschreibung der Vornembsten und bekantisten Stätte und Plätz in dem hochlöblichsten Churfürstenthum und March Brandenburg; und dem Hertzogtum Pomeren, zu sampt einem doppelten Anhang, 1 Vom Lande Preuen und Pomerellen 2 Von Lifflande unnd Selbige beruffenisten Orten." with engravings by Matthaeus Merian, 1652. In *Truck gegeben undt Verlegt durch Matthaei Merian Seel*, Erben, 1927. (also Frankfurter Kunstverein, 1959.) (also *Topographia Germaniae. Faksimile-Ausgabe. XIII: Brandenburg — Pommern — Preussen — Livland.* Bärenreiter-Verlag, 2005.) (also *Topographia Germaniae. — Brandenburg mit Preussen, Pommern, Liffland* Archiv Verlag.)

A.2 Sources Project

In 1982 I began a bibliographic project to trace and document the history of recreational mathematics. The effort concentrated on the lineage of individual puzzles. The previous section is built upon this work. In fact this entire book is composed of essays that document discoveries made during this decades long effort.

Unfortunately, the current state of the bibliography is that its parts were created with different formats and document tools and they have not coalesced. However, the parts are all accessible as

"Sources in Recreational Mathematics: An Annotated Bibliography," 8th preliminary edition, South Bank University, 2004. (Available online at the Puzzle Museum.) The level of detail and the array of problems far exceed what is covered in the volume in your hand. In this bibliography, as in this book, recreational is used in a fairly broad sense, but the more straightforward problems are omitted, to concentrate on those which "stimulate the curiosity" (as Montucla says).

In addition, recreational mathematics is certainly as diffuse as mathematics. Every main culture and many minor ones have contributed to the history. A glance at the Common References in the bibliography, or at almost any topic in the text, will reveal the diversity of sources which are relevant to this study. Much information arises from material outside the purview of the ordinary historian of mathematics — e.g. patents; articles in newspapers, popular magazines and minor journals; instruction leaflets; actual artifacts and even oral tradition.

Consequently, it is very difficult to determine the history of any recreational topic and the history given in popular books is often extremely dubious or even simply fanciful. For example, Nim, Tangrams, and Magic Squares are often traced back to China of about 2000 BCE. The oldest known reference to Nim is in America in 1903. Tangrams appear in China and Europe at essentially the same time, about 1800, though there are related puzzles in 18C Japan and in the Hellenistic world. Magic Squares seem to be genuinely a Chinese invention, but go back to perhaps a few centuries BCE and are not clearly described until about 80 CE. Because of the lack of a history of the field, results are frequently rediscovered.

Originally the idea was to produce a book (or books) of the original sources, translated into English, so people could read the original material. This bibliography began as the table of contents of such a book. This could have been an easy project, but it has become increasingly apparent that the history of most recreations is hardly known. Mathematical recreations are really the folklore of mathematics and the historical problems are similar to those of folklore. One might even say that mathematical recreations are the urban myths or the jokes or the campfire stories of mathematics. Consequently, it was decided that an annotated bibliography was the first necessity to make the history clearer. This bibliography alone has grown into a book, something like Dickson's History of the Theory of Numbers. Like that work, the present work divides the subject into a number of topics and treats them chronologically.

There have several preliminary editions of this work printed, with slightly varying titles. The first version of 4 July 1986 had 224 topics and 129 pages and was presented at the Strens Memorial Conference at the University of Calgary in July/August 1986. ... The eighth edition is 818 pages and there are more than 460 topics.

A fuller description of this project in 1984–1985 is given in my article "Some early sources in recreational mathematics," in: C. Hay *et al.*, eds.; *Mathematics from Manuscript to Print.* Oxford Univ. Press, 1988, pp. 195–208. Another description is in my article: "Recreational mathematics," in: *Encyclopedia of the History and Philosophy of the Mathematical Sciences.* ed. by I. Grattan-Guinness; Routledge & Kegan Paul, 1993, pp. 1568–1575.

A.3 Open Problems

There are a good many open problems throughout this book. There are other problems that are not treated here that are open. However, their histories are poorly understood. Here are some loose ends that have unknown antecedents.

Morris Games. The game known as Nine Men's Morris, Mill, Moule, Mühle, etc. is said to have been played at various places in the ancient world. A game board from 14 century BCE was found at Kurna, Egypt and boards from the first century occur at Mihintale, Ceylon. Are there other early examples? Are there any ancient written references?

Knight's Tours The earliest known example is in an Sanskrit manuscript poem Kāvyālaṅkāra, by Rudraṭa, c. 900, described in Murray's *A History of Chess.* Have any other early Indian versions been found?

Tangrams These became a fad in China c. 1810 and in Europe c. 1817.

In early 18th century Japan there was a similar puzzle with a different set of pieces. The only other early tangram-like puzzle seems to be the Loculus of Archimedes, known in the classical world from the third century BCE to at least the sixth century. It has 14 pieces and is rather more complex than the Tangrams. Some of the Archimedes sources are two 17th century Arabic manuscripts, so the Loculus may have been known to the Arabs and possibly they transmitted it to China, possibly passing through India. The recent recovery of the Archimedes Codex reveal that the loculus is discussed in it. The modern editor believes the text is describing how many ways the 14 pieces can be made into a square, making it the first combinatorial enumeration puzzle. Are there any Indian versions of the puzzle? Jerry Slocum has managed to track the predecessors of tangrams back considerably in China. See Figure A.1.

A 15th century ceiling panel in Seville is a square with some diagonal lines and looks like it could be used as a 10 piece tangram-like puzzle. Since geometric patterns and panelling are common features of Islamic art, are there any instances of such patterns being used for a tangram-like puzzle?

Casting Out Nines. This is often attributed to the Hindus, but there are some references to Greek special uses of it by St. Hippolytus (c. 200) and Iamblichus (c. 325) though these sources are unseen. Avicenna's attribution of it to the Hindus is said to be a dubious interpretation. It appears in al-Khwārizmī (c. 820) and al-Uqlīdisī (952) as well as in Āryabhaṭa II's *Mahāsiddhānta*.

The Chessboard Problem (which leads to $1+2+4+8+\cdots+2^{63}$) is often attributed to India, but the earliest known sources are Arabic: al-Yaqābī (c. 875) (described in Murray's *History of Chess*), al-Maṣūdīs *Meadows of Gold* (943), (which does not relate the series to the chessboard) and al-Bīrūnī's *Chronology of Ancient*

Figure A.1. *Chi Chiao Thu* [New Collection of Tangrams], 1823. It is a version of Sang-hsia-k'o's book of 1813.

Nations(1000). Murray cites a 9th or 10th century treatise on the problem by al-Missisī — does this exist? Is there any Indian or Persian material of relevance? See Figure A.2.

Weights. The use of $1, 2, 4, \ldots$ and $1, 3, 9, \ldots$ as weights occurs both in Fibonacci and in Ṭabarī (c. 1075), but there must be earlier examples.

Magic Squares The Indian and Arabic history of Magic Squares is quite confused. There are many Indian and Arabic references that remain unfound — indeed some apparently are not extant. A. N. Singh (*Proc. ICM*, 1936, pp. 275–276) refers to a first century order 4 square by Nāgārjuna, but it is not seen. *Bhairava Tantra* and *Siva Tāṇḍava Tantra* — *Dvivedi*, Introduction to Nārāyana Pandita's *Gaṇita kaumudī*, cites these as early sources for Magic Squares, but these remain a mystery. Varāhamihara II, *Pancasiddhantika* — Bag implies this says something about Magic Squares, but were not found in the English edition by Neugebauer and Pingree. This may be a confusion with his *Bṛhatsaṃhitā* which does have material on magic squares.

Liar Paradox Do the Liar or similar paradoxes occur in Arabic or Indian works?

Figure A.2. Columbia Algorism, c. 1350, F.47r.

Wire puzzles first seem to appear in the West in the late 19th century. They were and are popular in India and China, but except for the Chinese Rings, early sources are unknown.

Puzzle Rings are believed to be Middle Eastern in origin. Are there any sources? There is a 17th century example in the British Museum. We now know of a description by Pacioli, c. 1500.

Additional sources are welcome.